Autonomous Vehicles and Future Mobility

Autonomous Vehicles and Future Mobility

Edited by Jody Wheeler

Clanrye International
www.clanryeinternational.com

Clanrye International,
750 Third Avenue, 9th Floor,
New York, NY 10017, USA

ISBN: 978-1-64726-601-1

Cataloging-in-publication Data

Autonomous vehicles and future mobility / edited by Jody Wheeler.
p. cm.
Includes bibliographical references and index.
ISBN 978-1-64726-601-1
1. Automated vehicles. 2. Transportation--Forecasting. 3. Motor vehicles. I. Wheeler, Jody.
TL152.8 .A98 2023
629.2--dc23

For information on all Clanrye International publications
visit our website at www.clanryeinternational.com

CLANRYE INTERNATIONAL

Contents

Preface

An autonomous vehicle (AV) refers to a type of vehicle that is able to drive itself. It is a ground vehicle that can sense its surroundings and move safely with little or no human input. Self-driving cars use a range of sensors to sense their surroundings including GPS, inertial measurement units, radar, sonar, thermographic cameras and lidar. Sensory information is interpreted by advanced control systems to determine applicable signage, hurdles and proper navigation paths. They are also unaffected by human factors such as lack of attention and fatigue. This book provides significant information of autonomous vehicles to help develop a good understanding of its role in the future of mobility. Those in search of information to further their knowledge on automotive vehicles will be greatly assisted by it.

This book has been the outcome of endless efforts put in by authors and researchers on various issues and topics within the field. The book is a comprehensive collection of significant researches that are addressed in a variety of chapters. It will surely enhance the knowledge of the field among readers across the globe.

It gives us an immense pleasure to thank our researchers and authors for their efforts to submit their piece of writing before the deadlines. Finally in the end, I would like to thank my family and colleagues who have been a great source of inspiration and support.

Editor

1

Virtual Simulation Platform for Training Semi-Autonomous Robotic Vehicles' Operators

Cheng Siong Chin, Xionghu Zhong, Rongxin Cui, Chenguang Yang and Mohan Venkateshkumar

Abstract

This chapter covers the development of a virtual simulation platform for training a semi-autonomous robotic vehicle (SARV) operator via an open-source game engine called Unity3D. The SARV such as remotely operated vehicles (ROVs) is becoming increasingly popular in the maritime industry for risky jobs in inhospitable environments. The primary element in carrying out underwater missions in a hostile environment lies within the skills and experience of an ROV pilot. Training for ROV pilots is essential to prevent damage to expensive field equipment during the real operations. The proposed simulator differs from the existing simulators in the market is the use of modern game engine software to develop a "serious game" for ROV pilot trainee at much lower cost and shorter time-to-market. The results revealed that proposed virtual simulator can develop a high-fidelity virtual reality training for the underwater operation guided by classification society.

Keywords: autonomous robotic vehicle, Unity3D, remotely operated vehicle, simulation

1. Introduction

In recent years, the advancement of technology has dramatically improved the functions of remotely operate vehicle (ROV) [1–3] and autonomous underwater vehicle (AUV) [4] to handle the growing spectrum of underwater tasks [5]. Artificial Intelligence is becoming an increasingly common sight in automating machines to carry functions without the need for an actual operator. It will continue to take on a more observatory role [6–8]. Nowadays, ROVs and AUVs are commonly used in the maritime industry to carry out underwater tasks. These machines possess the capabilities to carry heavier loads to stay in deeper underwater for a longer duration than the human divers. These machines are operated from shore by a pilot making them very profitable and safer in the maritime industry. The challenge facing these ROV pilots is the ability to run the ROV with minimal information from the ROV feedback systems. As a result, the ROV pilot needs to be sufficiently skilled in maneuvering the vehicle in underwater.

With modern day technology, training simulators are developed to better equip ROV pilots with the necessary skills. Currently, the majority of ROV simulators available in the market are owned and distributed by companies that build ROVs. It is much cheaper to hone a pilot's competencies on a simulator than on the actual ROV as it creates room for the pilot to improve. It would better equip the pilots to deal with different underwater scenarios. Many ROV simulators are equipped with multiple scenes and a wide array of simulated sensors and equipment on the ROV.

This paper aims to develop a virtual simulation where ROV pilots can gain experience via a virtual environment using an open-source game development software to produce highly graphical visuals simulations for training purpose. The Unity3D game engine [8–10] was identified as a suitable development platform for the project due to its high graphics capabilities, built-in physics engine, well-documented manual, large online community and relatively mild learning curve in comparison with other game engines like the Unreal Engine [11] and CryEngine [12]. The details of designing a low-cost pilot simulator using a game engine are unique to this chapter.

The chapter is organized as follows. Section 2 presents a brief methodology for the training simulator. Section 3 discusses the virtual simulation development and followed by a conclusion.

2. Comparison of game engines

Game engines have the graphics and physics engines to build better and more realistic simulation. In this section, three commonly used open-source game engines are compared. For example, the commonly used software for game development is namely: Unity3D, Unreal 4 Engine, and CryEngine.

There exists few open source game engine software that provides excellent features and developing tools. Some of the common characters of a game engine are rendering, physics (2D and 3D rigid body), scripting, audio and animation. Depending on the requirements, these game engine software use traditional programming method that requires basic coding to high-level sandbox engine that provides "drag and drops" interface. The main objective is to simulate an ROV operation using high-level sandbox game engine for the ease of usability. The options are more toward the sandbox engine and the more common software for game development such as Unity3D, Unreal Engine and CryEngine. Although the software provides developers with a "drag and drop" interface, the features of each software somehow differ from one another. A brief comparison of various game engines is given below.

Unity3D was first released in 2005. It uses mostly JavaScript or C# or managed code tool chain that makes it simpler to support and develop new workflows and tools. It has large supporting communities that include the asset store for downloading different game characters, particle and sound effects. Due to its popularity, there exists a good educational material and large active user. However, the free version of Unity does not have Profiler that allows the programmer to optimize the game and check the time spent on rendering, and animation during the game. Unity3D supports around 21 platforms (PC, Web, Console, Mobile, etc.) as compared to Unreal Engine 4 supporting only around six platforms. Additionally, the 3D models in Unity3D can merely import as game assets into the software thus improving the efficiency of development.

Unreal Engine was first released in 1998. It provides developers with powerful tools such as access to full source code, simulates and immerse view, persona animation, and cascade visual effects. It is used in a custom workstation with better and optimized performance that implies higher cost and complexity. Unreal has much more extensive download than Unity3D as it requires visual studio for its programming environment and accepts only C++ development language. Unreal engine can produce high-quality graphics with advanced dynamic lightings making it a plus point for the game engine. However, the script used in Unreal Engine 4 can only be written in C++, which can be a drawback for beginners. Similar to Unity3D, Unreal Engine 4 has an asset store to download different game assets. However, the user community is not as large as Unity3D.

CryEngine started in 2002. It is another modern game engine that provides superb features that will create great realistic gameplay. With its pixel accurate displacement mapping, it allows developers to craft and modify a game as precise as possible. Its excellent graphics capabilities exceed those in Unity3D and Unreal Engine. However, a drawback from this game engine is that it requires a slightly higher learning curve before one can use the game engine efficiently and it may be harder for those with no game development background.

It is notable that these three game engines provide great features for the most development process. It can be quite subjective in the selection decision. Depending on the development objectives and requirements, one may pick Unity3D for its capabilities in developing 2D and 3D games, Unreal Engine for its powerful tools or CryEngine for its extreme graphics capabilities. Fortunately, these game engines are freely available for education and research except for Unity3D which requires Pro version for advanced features. On the other hand, the Unreal Engine and CryEngine require a slightly higher learning curve and posing difficulty for most beginners. Based on the following guidelines below, the free version of Unity3D that contains most of the functions will be used (at least for the beginning phase of the project) in developing a simulator for ROV pipeline tracking as it is easy to use, free for research purpose and the presence of wide user community. It may not be the best choice for every programmer, but during the development of the virtual reality simulator, there are no major problems encountered, and hence the choice was good enough.

- Able to communicate with external hardware
- Ease to program and use graphical user interface for controlling interaction and animating objects
- Able to process multimedia sensory data
- Free to use for education and research
- Able to hold multiple operating systems
- Able to support development with the strong developer community

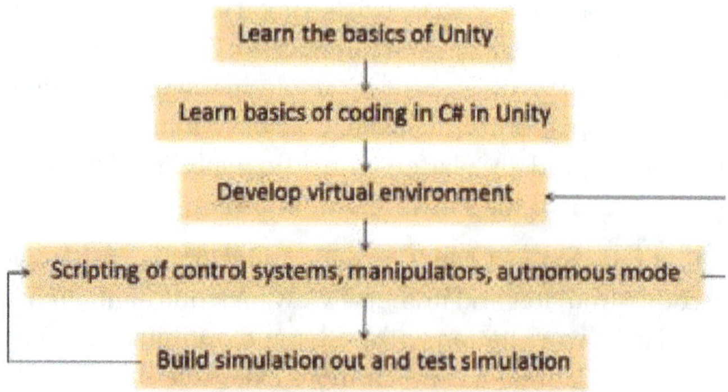

Figure 1. Steps to develop ROV pilot simulator.

The basic functions of the Unity3D Interface such as creating GameObjects, basics scripts and GameObjects for manipulation will be presented. The virtual environment and writing the scripts for the ROV's control system and manipulators will then follow. Assistance can be sought through the Unity3D manual found online or via Unity3D's online community and forums. As shown in **Figure 1**, the basic steps to develop the simulator can be seen. The simulator designed must be able to facilitate training of an ROV pilot. The environment of a typical pipeline inspection will be mimicked. After the leak is detected, the ROV will flash a bright red light indicating danger. The pilot will take control of the ROV and press the shutdown button of the BlowOut Preventer (BOP) to stop the leak and the flashing red light. After that, the pilot can continue the control of the ROV to carry out the autonomous inspection tasks. All controller inputs by the pilot are controlled via the joystick controller.

3. Virtual simulation development

Unity3D is a user-friendly off-the-shelf game development engine. The engine supports high visual graphics and physics to produce realistic 2D and 3D worlds, with readily developed assets available in the Unity3D Asset Store for users to download and import into their projects. A brief overall view of the software can be seen below.

3.1. Software interface

The Unity3D interface consists of several tabs and a toolbar (see **Figure 2**) are used to create all subsystems in the virtual simulation. The games are developed in the scene tab where GameObjects are added. These GameObjects are edited and programmed by combining various components such as textures, mesh, materials and scripts to make the GameObject behaves as required by the developer. As shown in **Figure 3**, the various components are placed into GameObjects under the virtual environment to produce the desired outcome.

3.2. Scene and model development

The scene developed in Unity3D is modeled after a subsea production system placed on the seabed 900 m below the surface. It consists of manifolds, pipelines, BOPs (BlowOut

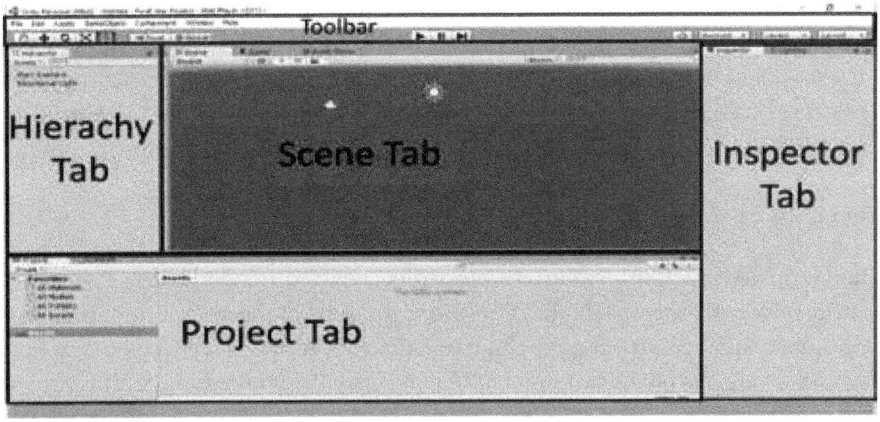

Figure 2. Unity3D interface.

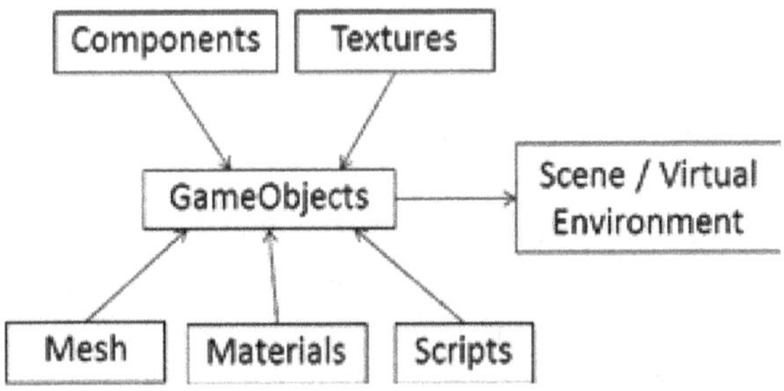

Figure 3. GameObject with various components.

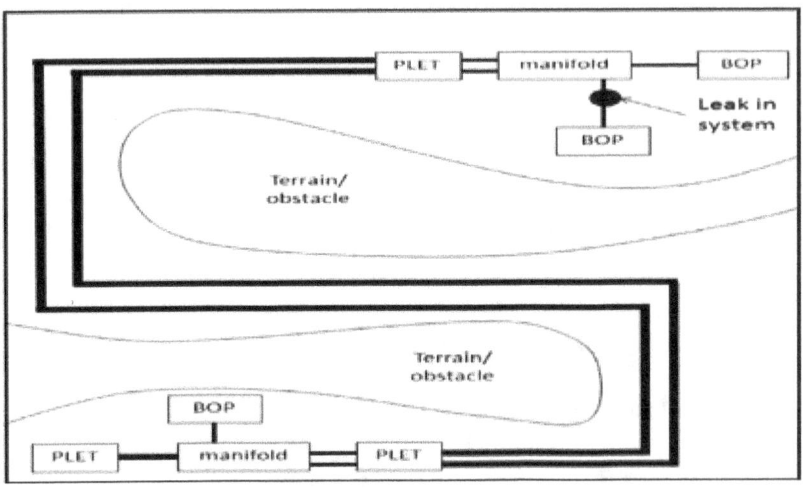

Figure 4. Map of simulated environment.

Preventer) and PLETs (Pipeline End Termination). A map of the simulated environment is shown in **Figure 4**. The scene was developed using models found online and crafted within Unity3D.

The ROV used in the simulation was modeled using an actual ROV used in the industry to fulfill requirement stated by the classification society DNV. It indicates that ROV models available in the simulation should be similar to the ROV in real life. In this proposed simulator, the TRV-M is a light work class ROV manufactured by Submersible Systems. The TRV-M in **Figure 5** can dive up to 1000 m deep that possesses the capacity to carry a payload up to 27 kg in any environments.

3.3. Autonomous operation

Automation is an area currently being developed in the maritime industry. The autonomous-mode component was added to the ROV in the simulator. In the simulation, the ROV begins in its autonomous inspection of the pipeline. It was achieved using a Unity3D function known as pathfinding where the object automatically computes the shortest path and moves itself to the desired point while avoiding the obstacles in the virtual environment. Using the pathfinding algorithm, multiple waypoints were added along the path of the pipeline to guide the ROV during the inspection.

3.4. Controller

To create a more realistic simulator, the control system for the ROV in the simulator should be similar to the control system of the ROV. A joystick controller or keyboard is one of the common controllers used in the maritime industry. The joystick controller will require multiple turning axes to accommodate all six degrees of motion similar to the movement of a vessel. The simulated ROV is capable of moving in six degrees of motion that is heave, sway, surge, yaw, pitch, and roll in **Figure 6**. As shown in **Figure 7**, the Logitech Extreme 3D Pro is used as the joystick controller to control the ROV movements and functions in the virtual environment. The ROV's controls are scripted and linked to the joystick via a Logitech Profiler. The various buttons and corresponding axis on the joystick are shown in **Table 1**.

3.5. User interface

The user interface (UI) is widely used in many control systems. It is designed to simplify the complicated tasks for the pilot to perform the functions. A well-designed UI can significantly reduce the cognitive load on the end user by merely displaying parameters and having a few buttons. Hence, it allows the user to control the system with focus on essential details [13]. The

Figure 5. TRV-M ROV.

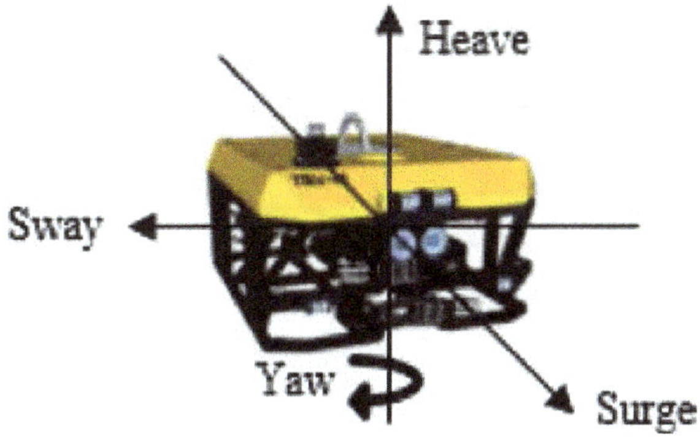

Figure 6. Model of TRV-M ROV with four degrees of freedom used in virtual environment.

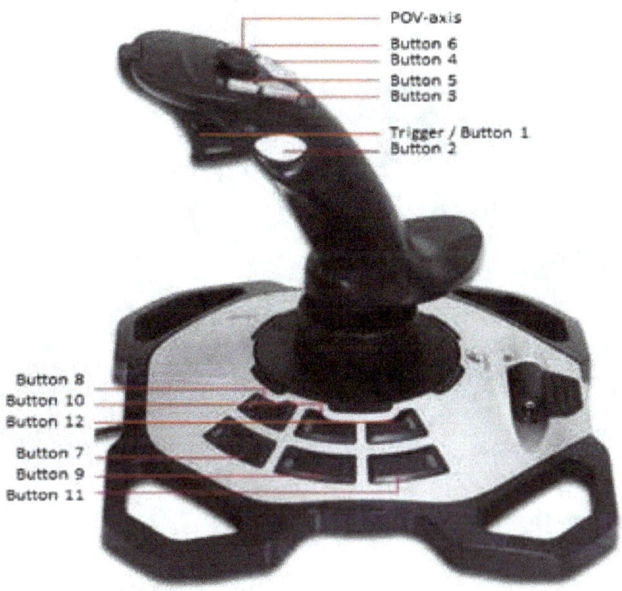

Figure 7. Logitech extreme 3D Pro joystick.

ROV function	Control key	
	Keyboard	**Joystick**
Move forward	W	Y-axis
Move backward	S	
Move left	A	X-axis
Move right	D	
Move down	Z	Button 11
Move up	X	Button 12
Yaw left	Q	Twist-axis
Yaw right	E	

ROV function	Control key	
	Keyboard	Joystick
Pitch up	I	POV-north
Pitch down	K	POV-south
Roll left	J	POV-west
Roll right	L	POV-east
Light	F	Button 9
Switch camera view	C	Trigger
Switch between modes	R	Button 3
Manipulator arms activate/deactivate	Tab	Button 2

Table 1. Player inputs for simulator.

Figure 8. User-interface parameters displayed.

UI designed in this project was created using a GameObject called Canvas, in Unity3D. The Canvas encompasses other objects within the User Interface (UI). **Figure 8** shows the UI displaying parameters such as latitude, longitude, depth, heading, speed, run time and small main menu to allow the pilot to restart the practice session.

4. Simulation results with understanding on maritime standards

By the classification standards in the maritime industry, all virtual simulators are assigned with a "Class" namely: Class A, B, C or Class S [14]. The class of a simulator is assigned based on the requirements of the simulator on the checklist provided by the classification society. The comparisons will be made with the standards proposed by DNV (Det Norske Veritas). The comparisons are not meant to be exhaustive but should provide adequate information for designing the virtual simulator.

From the results presented by the project, according to the standards of the classification society DNV, it can be seen that there are some criteria that the proposed Unity3D simulator could not meet. As seen in **Table 2**, items 1.1.19, 1.1.20 and 1.1.21 or any items related to the instructor, these requirements were not met because of the simulator was developed for a single purpose to train the ROV pilot in detecting a leak on a pipeline inspection. Therefore, it did not take into account an instructor to test the competency of the ROV pilot. Another observation in **Table 3** on the behavioral realism for items 2.1.1, 2.1.2, 2.1.5, 2.1.7 and 2.1.8, the class requirements were not fulfilled as the simulation does not encompass an underwater current acting on the objects in the scene. Simulation of actual underwater currents will affect the virtual objects and ROV to make the simulator more realistic. But one would require real information or data to implement it. It was not performed in this chapter. In **Table 4**, items 3.1.2, 3.1.4, 3.1.6 and 3.1.8 under the operating environment require a basic version of turbidity, sea state, underwater fog and camera in the virtual scene. However, some items were included such as camera and underwater lighting in the scene as shown in **Figure 9**. Others were not added due to the lack of actual specifications of these items.

Lastly, as seen in **Table 5**, casualty simulation was not met. The simulation did not have casualty simulation for the ROV. It did not include other equipment malfunctions besides the leakage that occurred in one of the pipelines (see **Figure 10**). In summary, the primary simulator produced has shown that Unity3D possesses the good capability to develop a more realistic virtual simulation for training purposes. Further developments of the simulator were performed to meet these class requirements. For example, a pathfinding algorithm (see **Figure 11**) was used to simulate the autonomous mode of the ROV.

In the virtual simulation, the pathfinding algorithm helps to build the autonomous mode feature in the ROV. Several waypoints were placed in the scene to better guide the ROV along its inspection path. However, using this method, the ROV would pass through the terrain to reach its next point. Few blocks were placed within the terrain to act as obstacles along the pathfinding. They appeared invisible in the simulator as their mesh renders can be switched off. However, the proposed guidance system would only be applicable if the simulated ROV carries out its autonomous functions at a fixed depth. A comprehensive way to create an autonomous guidance system for the ROV was to use a NavMesh algorithm as implemented in **Figure 11**. Any scene can easily be navigated by a GameObject with a NavMesh agent component. The NavMesh allows the simulated ROV to detect the size and height of objects in the scene that cannot pass through. The UI can be improved through the implementation of the User Interface where control buttons can be placed on the screen to allow the pilot to toggle them easily. The buttons on the GUI will serve to remind the pilot if a specific function is switched "on" or "off". As shown in **Figure 9**, the intended ROV GUI for the simulator with a button in the scene to toggle between the multiple camera views and front light on the ROV was implemented. As mentioned, the simulation of underwater currents will help the simulation to look more realistic for the pilot. With the physics supported by Unity3D, the ROV in the simulation can be subjected to these forces and moments caused by the underwater current. It will be implemented in the future works.

The simulation test results will focus on pipeline inspection tasks near to the seabed. The different parts of developing the simulation are integrated together with a scenario set up where there is a gas leakage in one of the jumpers in the top section of the subsea system. First, the ROV's

C1 physical realism		
Item	**Requirement**	**Fulfilled**
1.1.1	Displays and control configurations should be organized similarly to an actual ROV console.	No
ROV model		
1.1.2	ROV models available in the simulation should be similar to the ROV in real life.	Yes
1.1.3	An altimeter should be present to show the height of the ROV above objects.	No
1.1.4	The ROV should have a minimum of one manipulator. The manipulator shall have a similar outlook and behavior to that of it in real life.	Yes
Monitoring		
1.1.5	The simulated control panel shall have a similar outlook and encompass the information required by the pilot to operate the ROV.	Yes
1.1.6	Digital representations of the simulated ROVs thrust, lighting, depth and heading, date and time.	Yes
1.1.7	The digital sonar shall encompass a display window for navigation purposes as well as survey missions. It shall also provide detailed sonar images in consideration to the ROV's position and heading.	No
1.1.8	The tether to the ROV, if there is one, should be dynamic with a display showing the tension and length of the tether.	No
1.1.9	Navigational information should encompass depth, altitude, speed and heading of the ROV.	Yes
Control		
1.1.10	Control over thrust of ROV, hence allowing the pilot complete control of thruster.	Yes
1.1.11	Navigational control over the ROV's movement should be similar to those used for actual ROVs.	Yes
1.1.12	Individual manipulators should be manipulated via hardware controls.	Yes
1.1.13	The pilot should be able to switch between the various camera views available.	Yes
1.1.14	All panning and tilting units should be controllable by the pilot.	No
1.1.15	The pilot should be given a control function to dim the appropriate lights.	Yes
1.1.16	The pilot should be able to winch the tether fully at will.	No
1.1.17	An active or passive motion compensator.	No
1.1.18	A Launch and Recovery System (LRS) for Transportation Management System (TMS) operations, facilitating the docking and release of the ROV.	No
Instructor station		
1.1.19	A separate station with a 3D view should be implemented for the instructor.	No
1.1.20	Underwater visibility should be controllable by the instructor.	No
1.1.21	Underwater currents should be controllable by the instructor.	No

Table 2. Classification standards for physical realism.

task was to identify the location of the leakage, and second, the ROV will need to deactivate the whole system by pressing the switch on the subsea system. The data for the search operation were exported and plotted into a graph, and a video of the simulation was recorded.

C2 behavioral realism		
Item	**Requirement**	**Fulfilled**
2.1.1	A rigid body with 6 DOF (Degree Of Freedom) should be present in the ROV and have the ability resolve the various forces on objects within the simulation.	No
2.1.2	The simulated ROV should be able to resolve moments and forces acting on the ROV.	No
2.1.3	Deployment of the ROV from a vessel which follows the ROV on its own should be included.	No
2.1.4	Exercise areas, along with the various landmass, visuals, buoys tides, and depth data should be available when required to attain the desired training outcome.	No
2.1.5	The simulation should encompass multiple scenarios in dynamic environments with complete object interactions, using detailed dynamics, mechanisms and colliders.	No
2.1.6	The Sonar simulated and data attained from said simulated sonar should be similar to a sonar commonly used in real life.	No
2.1.7	It should be possible to simulate the effects of heave on various mechanical parts.	No
2.1.8	The ROV's tether should be responsive to the simulated environment.	No

Table 3. Classification standards for behavioral realism.

C3 operating environment		
Item	**Requirement**	**Fulfilled**
Visuals		
3.1.1	All visuals should have an engaging 3D environment with detail visuals.	Yes
3.1.2	Underwater visibility should controllable.	No
3.1.3	Lights should be adjustable by the pilot.	Yes
3.1.4	Pilot should have control over cameras allowing him to zoom, adjust and focus.	No
Environmental		
3.1.5	It should be possible to control the surge and speed of currents.	No
3.1.6	It should be possible to control sea state.	No
3.1.7	Interaction with sea floor should be realistic within simulation, e.g. clouding as a result of thrusters or suction.	No
3.1.8	Turbidity of the water should be controllable.	No

Table 4. Classification standards for operating environment.

As seen in **Figures 12** and **13**, the ROV starts at position (0, −793, 0 m), which is at the center of the subsea production layout. The ROV then rotates about its Y-axis in an anti-clockwise direction shown in **Figure 13**, where the ROV's heading is facing approximately 190° away from its global coordinate before it starts moving in X and Z-direction again. As shown in **Figure 14**, the ROV moves along at a constant depth of −793 m for the first 38 s before it moves up as the vehicle is too close to the seabed (to avoid collision). In **Figures 15** and **16**, the forward and lateral thrust (in N) produced by the ROV increases to reach the targeted location, i.e., to maneuver to the desired position. The thrusts reduce once the gas leakage

Figure 9. GUI with buttons to toggle between cameras and lighting.

C4 casualty simulation		
Item	**Requirement**	**Fulfilled**
4.1.1	It should have the capability to simulate impairment caused by severe strain on umbilical cord.	No
4.1.2	It should have the capability to simulate damages due to collision impact.	No
4.1.3	It should have the capability for the instructor to inject video display complications.	No
4.1.4	It should have the capability for the instructor to inject thruster control complications.	No
4.1.5	It should have the capability for the instructor to inject instrument display complications.	No
4.1.6	It should have the capability for the instructor to inject sonar display complications.	No

Table 5. Classification standards for casualty simulation.

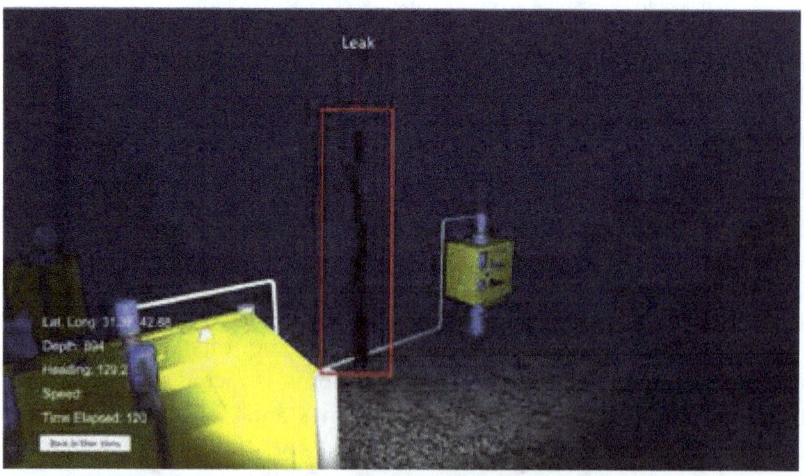

Figure 10. Leakage in pipeline.

Figure 11. Obstacles in the virtual scene.

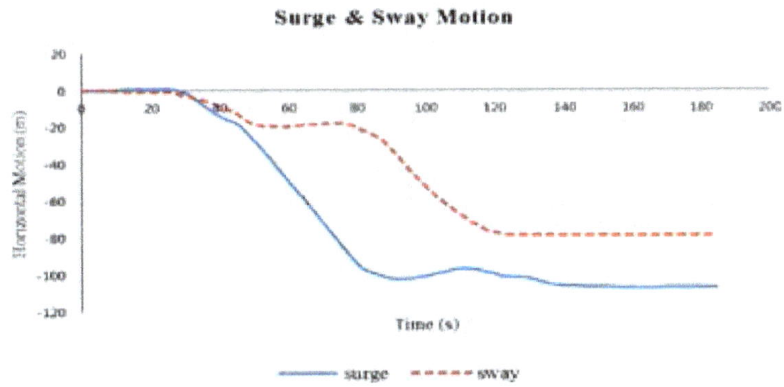

Figure 12. X and Y positions of ROV at the different time frame.

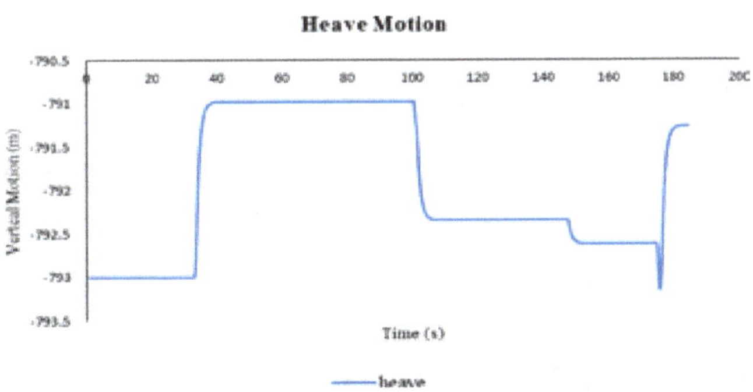

Figure 13. Z position of ROV at various time frames.

and subsea system are found. **Figures 12–14** show less fluctuating in the motion along the X, Z-axis, and yaw. The position of the ROV increases along the Z-axis while its position along the X-axis remains quite constant. The ROV makes a turn around the Y-axis (clockwise) to approximately 280° at 87 s where the position of the ROV increases along the X-axis and remain quite constant on the Z-axis. After approximately 140 s later, the ROV has reached the top section of the subsea system where the gas leakage is located in **Figure 10**.

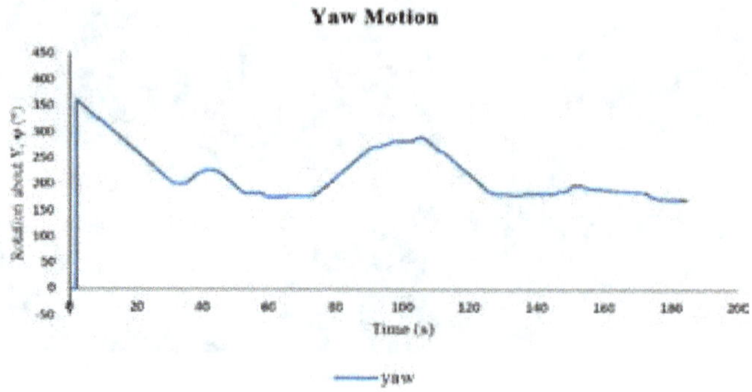

Figure 14. Yaw of ROV at various time frames.

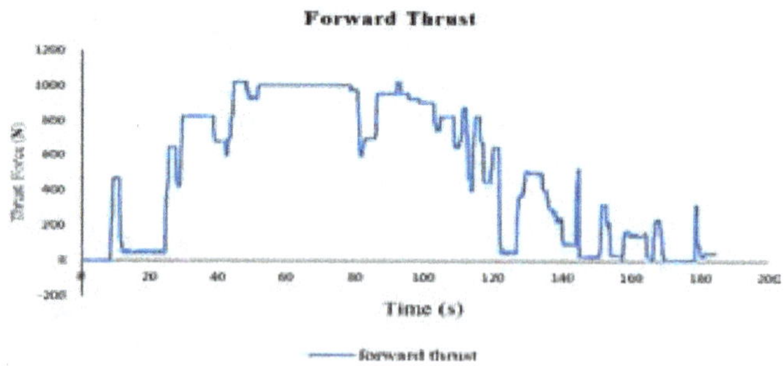

Figure 15. Forward thrust at different time frame.

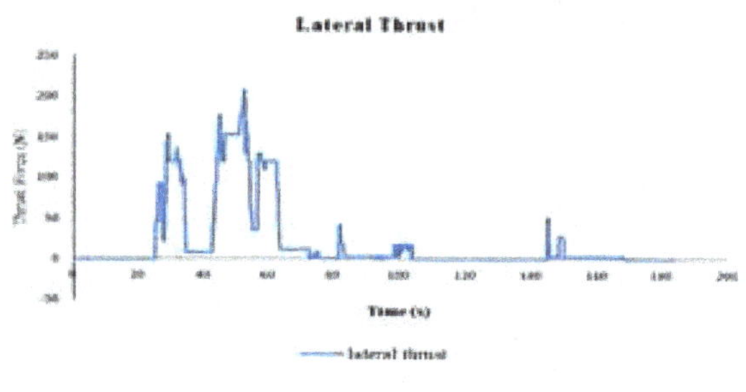

Figure 16. Lateral thrust at different time frame.

The left manipulator of the ROV was activated using a button on the GUI control panel to shut down the subsea system due to the leakage. The ROV can reach the targeted position and shut down system successfully after about 202 s. With the presence of the obstacles in the scene, the objects can interact with one another as though in the real environment. There was no obstacle being hit. The ROV pilot managed to find the gas leakage before reaching the targeted top section of the subsea system to deactivate the switch to prevent more leakages from happening. A sample of the simulation results obtained from the ROV simulator is summarized in **Table 6**.

Descriptions	Results
Total distance traveled in x direction	79.29 m
Total distance traveled in z direction	107.27 m
Final position on y-axis	−792.57 m
Final ROV is heading	169.75°
Total time was taken	202.18 s
Obstacle hits	0

Table 6. Simulated results obtained from ROV pilot simulator.

5. Conclusions

This chapter has successfully demonstrated the process of developing a virtual simulator using an open-source modern game engine software to develop "serious game" as a training system. By integrating different game objects using event-driven programming that is provided within the development software, it can be seen that modern game engine is capable of producing an appropriate level of accuracy in a fraction amount of time and at a low cost. Pilot training is useful for search and rescue operation, dynamic positioning of the drill string, subsea operation, and inspection tasks. Working at the simulator console, the trainees learn necessary flying skills and experience challenges of offshore operations before their first jobs. The virtual ROV's simulator creates a powerful and an efficient project preview and assessment tool for training. The low-cost ROV's pilot simulator can simulate live inputs from the vessel, rig to position the drill string and co-ordinates work operations efficiently in an offshore environment. It dramatically enhances the current training capability, workplace safety in an uncertain environment and reduces maintenance cost as the present cost of a complete ROV's pilot simulator is quite expensive to use and maintain.

Further research can be made to examine whether a pilot candidate who experienced the training in a simulated environment will perform better than those who do not have prior training. The design of different controllers can be implemented to add robustness to the ROV operation in underwater. Further development could be shown to improve on the simulator to suit the requirements of the classification society and specifications. It will be necessary as ROVs are used in many underwater tasks with more autonomy in its operations. However, it is still not possible to replace the decision-making process of a human operator in the loop.

Acknowledgements

The authors would like to thank Newcastle University for supporting the project.

Author details

Cheng Siong Chin[1*], Xionghu Zhong[2], Rongxin Cui[3], Chenguang Yang[4] and Mohan Venkateshkumar[5]

*Address all correspondence to: cheng.chin@ncl.ac.uk

1 Newcastle University, Singapore, Singapore

2 Visenti Pte Ltd, Singapore, Singapore

3 School of Electronic and Control Engineering, Northwestern Polytechnical University, Chang'an University, Xi'an, China

4 Zienkiewicz Centre for Computational Engineering, Swansea University, Swansea, UK

5 Department of EEE, Aarupadai Veedu Institute of Technology, Chennai, Tamil Nadu, India

References

[1] Christ R, Wernli SR. The ROV Manual, Second Edition: A User Guide for Remotely Operated Vehicles. Second ed. Waltham, USA: Elsevier; 2014

[2] Chin CS, Lau MWS, Low E. Supervisory cascaded controllers design: Experiment test on a remotely-operated vehicle. Proceedings of the Institution of Mechanical Engineers: Journal of Mechanical Engineering Science. 2010;**225**(3):584-603

[3] Chin CS, Lau MWS, Low E, Seet GG. Software for modelling and simulation of a remotely operated vehicle. International Journal of Simulation Modeling. 2006;**5**(3):114-125

[4] Cui R, Chen L, Yang C, Chen M. Extended state observer-based integral sliding mode control for an underwater robot with unknown disturbances and uncertain nonlinearities. IEEE Transactions on Industrial Electronics. 2017;**64**(8):6785-6795

[5] Cui R, Yang C, Li Y, Sharma S. Adaptive neural network control of AUVs with control input nonlinearities using reinforcement learning. IEEE Transactions on Systems, Man, and Cybernetics: Systems. 2017;**47**(6):1019-1029

[6] Schjolberg I, Utne IB. Towards autonomy in ROV operations. IFAC-PapersOnLine. 2015; **48**(2):183-188

[7] Chin CS, Lin WP. Robust genetic algorithm and fuzzy inference mechanism embedded in sliding-mode controller for uncertain underwater robot. IEEE/ASME Transactions on Mechatronics. 2018;**32**(2):655-666

[8] Xie J. Research on key technologies base Unity3D game engine. In: 7th International Conference on Computer Science & Education (ICCSE); Melbourne, Australia: 2012. pp. 695-699

[9] Chin CS, Kamsani NB, Zhong XH, Cui R, Yang C. Unity3D serious game engine for high fidelity virtual reality training of remotely-operated vehicle pilot. In: 10th International Conference on Modelling, Identification and Control, Guiyang, China; 2-4 July 2018

[10] Bartneck C, Soucy M, Fleuret K, Sandoval EB. The robot engine—Making The Unity 3D game engine work for HRI. In: Proceedings of the IEEE International Symposium on Robot and Human Interactive Communication (RO-MAN2015), Kobe; 2015. pp. 431-437

[11] Altabel, Altabel Group's Blog Unreal Engine 4, Unity, Cry Engine: What to choose? Available from: https://altabel.wordpress.com/2015/01/22/unreal-engine-4-unity-cry-engine-what-to-choose/ [Assessed: 14 May 2018]

[12] Juarez A, Schonenberg W, Bartneck C. Implementing a low-cost CAVE system using the CryEngine2. Entertainment Computing. 2010;**1**(3-4):157-164

[13] Candeloro M, Sorensen AJ, Longhi S, Dukan F. Observers for dynamic positioning of ROVs with experimental results. IFAC Proceedings Volumes. 2012;**45**(27):85-90

[14] Veritas DN. Standard for Certification No. 2.14 Maritime Simulator Systems. Det Norske Veritas (DNV) Standards for Certification, Høvik, Norway. 2011. pp. 75-78

2

Sky-Farmers: Applications of Unmanned Aerial Vehicles (UAV) in Agriculture

Chika Yinka-Banjo and Olasupo Ajayi

Abstract

Unmanned aerial vehicles (UAVs) are unpiloted flying robots. The term UAVs broadly encompasses drones, micro-, and nanoair/aerial vehicles. UAVs are largely made up of a main control unit, mounted with one or more fans or propulsion system to lift and push them through the air. Though initially developed and used by the military, UAVs are now used in surveillance, disaster management, firefighting, border-patrol, and courier services. In this chapter, applications of UAVs in agriculture are of particular interest with major focus on their uses in livestock and crop farming. This chapter discusses the different types of UAVs, their application in pest control, crop irrigation, health monitoring, animal mustering, geo-fencing, and other agriculture-related activities. Beyond applications, the advantages and potential benefits of UAVs in agriculture are also presented alongside discussions on business-related challenges and other open challenges that hinder the wide-spread adaptation of UAVs in agriculture.

Keywords: agriculture, crop production, farming, livestock, unmanned aerial vehicles

1. Introduction

An Unmanned Aerial Vehicle (UAV) is a type of aircraft that operates without a human pilot on-board. There are different types of UAVs employed for different purposes. Originally, the technology was employed by the military for anti-aircraft target practice, intelligence gathering and surveillance of some enemy territories. The technology has however grown beyond its initial purpose and in recent years has gained prominence in different spheres of human endeavor. Advancements in technology has allowed for the increased adapta tion of unmanned aerial vehicles for various purposes. Without an on-board pilot, UAVs are controlled either remotely by a pilot at a ground station or autonomously, steered by a pre-programmed flight plan.

There is a huge potential for the application of UAVs in Agriculture. One such application is in accurate and evidence-based forecasting of farm produce using spatial data collected by the UAV. UAVs also allow farmers to observe their fields from the sky. This sky-view can reveal many issues on the farm, common among which is irrigation related problems, soil variations, fungal and pest infestations. Further information relating to water access, changing climate, wind, soil quality, the presence of weeds and insects, variable growing seasons, and more can all be monitored with UAVs. From a livestock perspective, UAVs are being used to perform head counts, monitoring animals and also studying eating habits and health related patterns. Utilizing the information gathered, farmers can provide fast and efficient solutions to detected problems and issues, make better management decisions, improve farm productivity, and ultimately generate higher profit. In this chapter, various applications of UAVs in Agriculture are discussed both in commercial livestock farming and crop farming. This chapter also presents some of the open challenges to the application of UAVs in Agriculture.

Immediately following this introduction is a discussion of the various types of UAVs which is done in Section 2. This is followed by the applications of UAVs in crop farming and in livestock in sections 3 and 4 respectively. Advantages of UAVs and corresponding challenges are discussed in Section 5, the chapter ends with the 6th and concluding section.

2. Types of UAVs

UAVs can be classified based on usage, with some being used for photography, aerial mapping, surveillance, cinematography etc. However, a better classification can be made based on their feature sets. Vroegindeweij, et al. in their paper [1], presented an overview of the different types of UAVs applied in Agriculture and categorized them into three main groups – fixed-wing, Vertical Take Off and Landing (VTOL) and bird/insect. The authors identified the VTOL with its agility, great maneuverability and hovering ability as best suited for Agricultural application. In [2], the authors however argued in favor of the fixed-wing UAVs, stating that their long flight time and speed makes them better suited in comparison to the VTOL, which have comparatively shorter flight time and slow speed. In other works, authors have argued in favor of unmanned helicopters such as the monocopter or single-rotor UAV [3, 4]. These types of UAVs have long flight time, can fly at different altitudes and have good hovering abilities. However, they are much more complex to fly. A comprehensive survey of various UAVs was also done in [5]. From these literatures, four major types of UAVs are identified, which are:

1. Multi-rotor UAVs
2. Fixed-wing UAVs
3. Single-rotor Helicopter
4. Fixed-wing-multi-rotor Hybrid UAVs

2.1. Multi-rotor UAVs

These are the most common type of UAV, evident by their wide popularity among professionals and hobbyists alike. They find applications in photography, aerial video surveillance, recreational sports and games etc. They are the easiest to manufacture and also the cheapest type of UAV. Multi-rotor UAVs are further classified based on the number of rotors on the platform. There are those with three rotors called tricopter, with four rotors called quadcopter, with six rotors called hexacopters and those with eight rotors called octocopter. Flying a multi-rotor UAV does not require exceptional skill unlike the other types of UAVs.

Multi-rotor UAVs though cheap and easy to manufacture have a few drawbacks which include: limited flying time, endurance and speed. They can only sustain an average flying time of between 20 and 30 minutes. This is because a large percentage of their energy is expended fighting gravity and wind to remain stable in the air. **Figure 1** shows an octocopter used for precision spraying of liquid pesticides and herbicides.

2.2. Fixed-wing UAVs

These types of UAVs have wings similar to normal aircrafts. Unlike the Multi-Rotor UAVs, they do not exert a lot of energy to stay afloat in the air, hence able to fly longer; having average flight times of over an hour. Longer flight time makes them most ideal for long distance operations. However, they cannot hover on a spot and are thus not suitable for aerial photography. Furthermore, they are more expensive and require exceptional flying skill to operate. **Figure 2** shows a sample fixed wing UAV used for capturing images across large acres of farmland.

2.3. Single-rotor UAVs

Single rotor UAVs are also called monocopters and look very much like helicopters in design and structure. Though they are called single rotor UAVs, they actually have two rotors - a large on top and a smaller one at the tail. The bigger rotor is for lift while the smaller is

Figure 1. Multi-rotor UAV [6].

Figure 2. A fixed-wing UAV [7].

used for control. They have significantly longer flying time than their multi-rotor counterpart, as they are often powered by gas engines. These UAVs are also highly maneuverable and much more efficient than the multi-rotor types. Similar to the multi-rotor, they are also able to hover, hence useful for aerial photography and precision spraying. Despite these beneficial attributes, they come with higher operational risks as the large sized rotor blades usually pose a risk which is mostly fatal in nature. Like the fixed wing UAVs, these also require special flying training. **Figure 3** shows a sample single-rotor UAV.

2.4. Fixed-wing-multi-rotor hybrid UAVs

These types of UAV combine features of the fixed-wing and the multi-rotor UAVs, with the hybridization gives these UAVs a best-of-both-worlds feature set. They are able to perform vertical take-off and land (VTOL) as well as hovering in place like the multi-rotor and single-rotor. Similar to the fixed-wing and single-rotor UAVs, these also benefits from long flight-time, but can stay in flight for much longer. **Figure 4** shows an image of one such UAV that is versatile enough to be used for image capturing, surveillance as well as precision spraying.

Figure 3. A single-rotor UAV [8].

Though these are the four common types of UAVs, there is a unique type of UAV called the Flexible Membrane Wing (FMW) UAV [10]. The FMW has wings made from flexible membrane material, with the advantages of this being easy of storage (as the wings can simply be folded up) and better control and maneuverability in windy conditions (as the flexible wings dynamically adjust to cater for wind preventing "adaptive washout"). The FMW is a niche UAV, targeted flying in harsh and windy conditions. Flexible membrane also implies lighter weight and by extension the possibility of carrying larger payloads.

2.5. Comparison of various UAVs

Figure 5 shows a comparison of the four different types of UAVs based on their average weights, payload size and flight time; **Table 1** on the other hand summarizes their in-flight specifications. For each category, a model UAV was selected. The values shown were obtained from the respective manufacturer documentation and/or operator's manual of each product. On **Table 1**, the advantage of the hybridization can clearly be seen, as it resulted in

Figure 4. A hybrid fixed-wing-multi-rotor UAV [9].

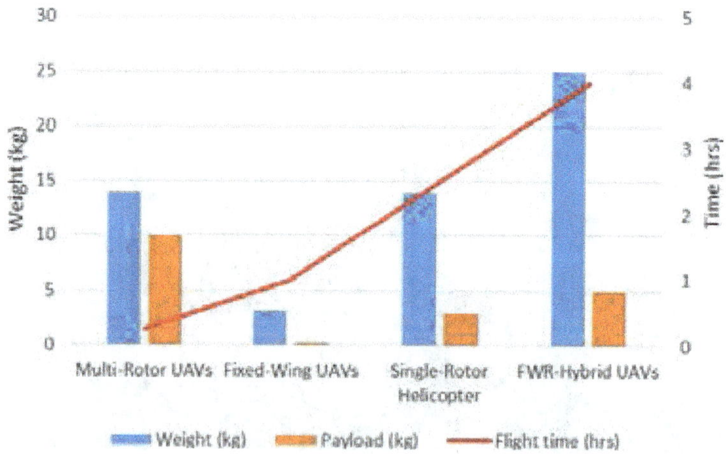

Figure 5. UAV weight and payload vs. flight time.

UAV type	Altitude (km)	Avg. control range (km)	Avg. airspeed (m/s)
Multi-rotor UAVs (DJI Agras MG-1P [6])	2	3–5	7
Fixed-wing UAVs (AgEagle RX60 [7])	0.125	2	18.8
Single-rotor (Alpha 800 [8])	3	30	15.2
Fixed-wing-multi-rotor hybrid UAVs (Jump 20 [9])	4	500–1000	30

Table 1. Feature-based comparison of UAVs.

higher flying altitude, wider control range, increased speed and longer flight time compared to the other UAV types.

3. UAVs in plant/crop farming

According to Massachusetts Institute of Technology (MIT), UAV technology will give the Agriculture industry a high-technology makeover, with planning and strategy based on real-time data gathering and processing. PwC put a $32.4 billion valuation on the UAV-powered Agriculture solutions market [11]. The application of UAV technology in Agriculture has become increasingly necessary with the increase in global population and the resultant pressure on agricultural consumption. The ever growing international population is not proportionately matched with crop growth; hence, there is a growing concern about food sustainability. In a bid to tackle this challenge, farmers around the globe have had to adapt modern and automated solutions in order to keep up with the agricultural needs of the world population that is in constant flux. UAVs are one such technology that could help improve crop yield. A number of UAV application areas are presented in the following subsections.

3.1. Soil and field analysis

The use of UAVs for soil information sourcing is helpful at the early start of a crop cycle. The data collected helps in early soil analysis, and is also useful in planning seed planting patterns. These data can also assists the farmer in making irrigation plans as well as determining the quantity of fertilizer needed on the soil or field after planting. Using a data-driven approach, the farmers can improve the overall yield quantity of agricultural produce, while significantly saving on fertilizers and pesticides. All these are made possible through the analysis of remote images captured with UAV. UAV imagery also has a huge potential in designing site-specific weed control treatments. With the high resolution images, farmers can quickly and precisely spot weeds almost immediately they spring up and apply minimal pesticide to contain them. The authors in [12] developed an Object-Based Image Analysis (OBIA) on a series of UAV images using six-band multi-spectral cameras on a maize field in Spain. While in [13] the technical specifications and configuration of a UAV which could be used to capture remote images for Early Season Site-Specific Weed Management (ESSWM) were given. The study also evaluated the image spatial and spectral properties necessary for weed seedling discrimination. They deployed an UAV equipped with multi spectral cameras and analyzed the technical specifications and configuration of the UAV to generate images at different attitudes; with the high spectral resolution required for the detection and location of weed seedlings in a sunflower field. The result of the study can be of help in the selection of an adequate sensor and configuration of the flight mission for ESSWM.

3.2. Planting

Planting crops is a costly and cumbersome endeavor that has traditionally requires a lot of manpower. UAVs have simplified crop planting for farmers, with their abilities to cover large acres of land within a short period with utmost precision and accuracy. Today's high-end UAV farming technology offers UAV-powered planting techniques that reduce planting costs by up to 85%. The reduction in planting costs is a result of the UAV's capability of performing multiple tasks at the same time.

UAVs have become increasingly popular in recent years in agricultural research applications. They have been found to have capabilities of acquiring images with high spatial and temporal resolutions in Agriculture. Reference [14] evaluated the performance of a UAV-based remote sensing system for quantification of crop growth parameters of six sorghum hybrids. Factors such as Leaf Area Index (LAI), fractional vegetation (fc) and yields were considered. The evaluation was carried out using a fixed-wing UAV, equipped with a multi spectral sensor to collect images during the 2016 growing season with flight missions carried out 50 days after planting. The flight missions provided data covering the different growth periods of the sorghum hybrids. The authors inferred that high resolution images acquired using UAV can be effectively utilized for in-season data collection from the field. The results obtained proved the relationship between Normalized Difference Vegetation Index (NDVI) and LAI, and between NDVI and fc. It was thus possible to determine/estimate LAI and fc from UAV derived NDVI values. It was shown also that imagery taken at flowing stage could be better indicator of yield, rather than NDVI obtained at earlier growth stage of sorghum crop. Furthermore, it was also established that early season NDVI measurement is useful index for estimating plant population density of sorghum.

The authors in [15] sought to develop a novel method to quantify the distance between maize plants at field scale using an UAV. The distance between roots and plants are essential in determining the final grain yield in row cops. An UAV-based image algorithm was developed to calculate maize plant distances. Knowledge of the exact number of plants per square meter is essential and helps to improve yields by deducing the fertilizer and pesticide application to match plant demand. Determining plant population is essential for several other processes such as soil-to-plant balance, nutrient recycling and resource use efficiency. The study demonstrated the possibility of quantifying the distance between maize plants and provided an innovative approach to quantify plant-to-plant variability and by extension crop yield estimates.

3.3. Crop and spot spraying

Crop spraying is usually a tough and onerous task for farmers and agricultural production companies. It involves covering extremely large expanses of land comprehensively to ensure proper growth of crops. Agricultural UAVs have simplified crop spraying for farmers; as they can cover large expanse of land within a very short time interval. Using sensors, UAVs can automatically adjust their height when spraying across uneven fields. This improves the spraying accuracy and conserves resources. The advantages of using UAVs for crop spraying include: time and cost savings for the farmer, efficient spraying as both the plants and the soil below can be reached, and protecting farmers from prolonged exposure to potentially harmful chemicals that are hitherto associated with manual spraying. Agricultural UAVs utilize state-of-the-art topographical scanning techniques to dispense the optimal amount of fluid required for proper crop growth. This ensures even coverage with limited wastage. Lv et al. [16] demonstrated the practicability of infrared thermal imaging in evaluating the droplet deposition in the field of aerial spraying. In the study, the effect of UAV flight speed on the spray droplets was investigated and the variable spray test was conducted by a UAV simulation platform, with airborne spray system under controllable environment. Several conclusions were drawn from the study among which were that deposition density decreases with the flight speed and droplet diameter (i.e. the distribution uniformity of particle size) decreases with an increased flight speed resulting in the worse uniformity of the sprayed droplets. The authors therefore provided a theoretical support for optimizing the spraying parameters of plant protection UAV, aimed at improving plant yield.

Spot spraying is similar to crop spraying but targets weeds. With the use of high resolution cameras, the UAV can identify weeds and precisely spray a jet of herbicide. Spot spraying can save up to 90% on chemical herbicides. Numerous research works [17–19] have been done in determining the efficacy of UAVs for spot spraying. Some factors considered were balancing UAV altitude and speed with spraying height and accuracy as well as droplet sizes, spray pressure and the possible effects of the UAVs' propeller(s) airflow direction.

3.4. Crop monitoring

A combination of large farm fields and low efficiency in crop monitoring system are some of the greatest farming challenges. The challenge of monitoring is further aggravated by unpredictable weather conditions, which drive up risk and field maintenance costs. An agricultural UAV helps the farmer overcome some of these challenges. UAVs with thermal imaging cameras enable the farmer to monitor his farm. The farmer can check the state of crops in the farm, as well as areas that need urgent attention. The result is improved yield and greater profit. [20] demonstrated the possibility of generating quantitative mapping products such as crop stress maps from UAV images and highlighted the value of UAV remote sensory when applied in precision Agriculture. The study applied a single-rotor UAV (monocopter), equipped with multiple spectral cameras, and then developed a framework to process the UAV images and generate mosaic images which can be aligned with maps for GIS integration at a later stage.

3.5. Irrigation

Agricultural UAVs fitted with thermal imaging cameras have the capability to providing tremendous insights into specific troubled areas in the farm. Using the thermal cameras, the farmers are able to determine areas with low soil moisture, pinpoint crops that are dehydrated, locate areas that are water-logged and in general have a sense of the overall health status of crops in the field. Such precise and specific monitoring were either not possible with traditional farming, inefficiently done or extremely expensive as experts have to be contracted to carry out the task and proffer adequate solutions. UAVs now give the farmers that ability to do these themselves. In [21], the authors carried out a study on vineyard water status variability by thermal and multispectral imagery using an UAV. Assessment of the water status variability of a commercial rain-fed Tempranillo vineyard was done, and concluded that an UAV can be used to assess vine water status, and to map within vineyard variability which could be useful for irrigation practices. The work done in [22] focused on the application of thermal remote sensory in precision Agriculture, and some of the concerns relating to its application. Gonzalez-Dugo et al. [23] further dealt with the assessment of heterogeneity in water status in a commercial orchard as a prerequisite for precision irrigation margent. High resolution airborne thermal imagery was employed. A UAV with thermal camera on board was flown three times during the day over a commercial orchard; and the indicators derived from the thermal imagery described the spatial variability in crop water status and thus allows the mapping of an orchard on a tree by tree basis. It therefore becomes a valuable tool for water management in precision Agriculture.

3.6. Health assessment

Farm health assessment is crucial for detecting fungal and bacterial diseases on the farm. By scanning a crop using both visible and near-infrared light, UAV-carried devices can detect temporal and spatial reflectance variations and associate it to the farms health for early interventions, which ultimately saves the entire farm. These two possibilities increase a plant's ability to overcome disease. And in the case of crop failure, the farmer will be able to document losses more efficiently for insurance claims. UAVs offer new and modern methods of accurately monitoring and assessing pest damage needs to be investigated. The authors in [24] explored the combination of UAVs, remote sensory and machine learning techniques as a promising technology to address the problem of agricultural pests in farmlands. UAV platform was deployed over a sorghum crop in South-East Queensland, Australia, to collect high resolution RGB images of certain areas which were severely damaged by white grub pest. An image processing pipeline was implemented prior to image analysis. The study demonstrates how UAV-based remote sensitivity and machine learning could be used to achieve biosecurity surveillance and pest management. The work presented in [25] also corroborated the use of UAV in crop health assessment, and outlined the benefits of deploying UAV remote sensing over the traditional methods. They developed a method that can quickly monitor crop pest, based on UAV remote sensing, which was deployed for inspection pests in Baiyangdian agricultural zone during the growth season. An improved SIFT Algorithm was adapted for image matching and mosaic with good result. The method adopted by [24] was used to check the status information of crop pest. Similarly, in the work done by Yinka-Banjo et al. [26], the authors proposed the use of UAVs for bird control in farmlands. Their solution combined the use of autonomous vehicles with bird scare tactics. The combination was reported to be more efficient than the traditional human-based manual approaches.

4. UAVs in livestock farming

Livestock farming is the act of rearing animals for food and/or other uses such as medicine, leather, fur and fertilizer. The authors in [27, 28] showed that traditionally Livestock Production Systems (LPS) were grouped into three major classes, namely: livestock production integrated with crop, land based and agro-ecological. They further sub-divided LPS into 11 groups – solely livestock production, temperate and tropical highlands grassland-based, arid and subtropics grassland, humid and subtropical mixed-farming based, temperate and tropical highlands rain-fed mixed farming, humid and subtropics rain-fed mixed, temperate and highlands irrigated mixed farming, humid and subtropics irrigated mixed farming, arid and subtropics irrigated mixed farming, landless monogastrics and landless ruminant farming. Similarly, in [29], the authors reviewed five (5) types of livestock production systems in tropical areas based on factors such as agro-ecological zones, animal type, function and management. The identified classes were Pastoral Range, Crop-livestock (low and highlands), Ranching and landless.

With respect to livestock, sheep and goats are the most farmed animals, followed by cattle. **Table 2** shows a numerical distribution of global livestock produce adapted from [30].

Livestock farming as with other aspects of Agriculture can be monotonous and laborious. Humans are however not well suited to such task over a prolonged period of time. Machines therefore can find practically applications in this arm of Agriculture, as they are designed to perform repeatable tasks, faster and possibly more efficiently (over a long period of time) than humans can. UAVs are therefore no exceptions and have found practical applications in livestock farming. Applications of UAVs in livestock farming are discussed as follows:

Animal type and/or produce	Number/quantity (10^6)
Animal	
Sheep and goats	1777
Cattle and buffaloes	1526
Animal produce	
Milk	594.4
Pork	95.2
Poultry meat	73.7
Beef	60.7
Eggs	58.9
Mutton	11.9

Table 2. Global livestock produce.

4.1. Livestock censors

To further put **Table 2** in perspective, according to the National Development Agency of South Africa, there were over 13 million units of cattle, 30 million sheep and 6.6 million goats and 1.6 million pigs bred in each province annually between up on to 2003. The figures are even significantly higher in European countries according to Eurostat. These are staggering numbers, hence monitoring and daily head counts of these large number of animals can be challenging. UAVs can thus find application here and be used to perform headcounts of livestock across these large grazing areas [31–33]. Animal counting can be done either by using image recognition [31] or using heat detecting infra-red cameras [34]. For image processing, Convolutional Neural Network (CNN) has emerged in recent times as the most widely used [35]. In large grazing areas, the UAVs can also be used to detect and count the number of animals present. In most of these works, the UAVs fly across the field, and counting the number of animals present. In the work done by [33] however, the authors proposed an approach, wherein the number of goats are counted and tracked using fewer numbers of pictures, sometimes only one. The authors reported 73% count accuracy and 78% tracking accuracy.

In contrast, in their book [34], the authors reviewed numerous methods of performing thermal imaging for monitoring animals in the wild. Among many other factors, the authors argued that thermal imagining is not dependent on time of the day unlike image processing. This therefore provides a unique opportunity to observe animals in their natural habitats without causing disturbances – which can lead to dispersion and possibly double or inaccurate counts.

4.2. Animal health

Beyond counting, research work is underway at the Texas A&M University, to investigate the use of infra-red cameras mounted on UAVs to monitor the health of animals. The research is based on the premise that, animals with fever tend to have heightened temperature. This can easily be detected by the UAV and appropriate medication can be administered [36]. Similar research targeted at monitoring health has also been carried out in [37]. **Figure 6** shows a sample heat map of a herd of cattle captured by a UAV.

4.3. Monitoring and identification

On an individual levels, animals can be tagged with RFIDs or similar sensors and can then be monitored using UAVs. With this, farmers can effectively monitor the movements and feeding behavior of a specific animal [38]. This has also been extensively used in monitoring endangered animals, raised in captivity and released into the wild. Similar to the two application areas discussed above, the identification can be carried out using UAVs fitted with normal cameras or IR cameras (which detect heat emissions from the animal) or RFID readers.

A major challenge to the application of RFID is that passive RFID tags have very short range, hence might be difficult to use. Potential solutions might include:

1. Painting QR codes on cattle, which the cameras fitted on the drone, can simply scan in order to identify the animal.
2. The use of a relay drone, such as the RFly being researched at MIT [39]. The RFly acts as a relay between the RFID tags and the reader. Using RFly, the authors recorded up to 50 meter range extension for passive RFID tags.

These technologies can be borrowed and used for animal identification and monitoring in Agriculture.

Figure 7 shows a potential use case of UAVs and RFID tags in animal identification.

4.4. Aerial mustering

Mustering is the process of using aircraft to locate and gather animals across a large span of land. Dogs (sheep dogs) and human on horses (cow boys) or motorcycles have traditionally been used to direct livestock along specific paths. For larger expanse of land, small sized helicopters are used. These helicopters are often piloted by one person and have highly

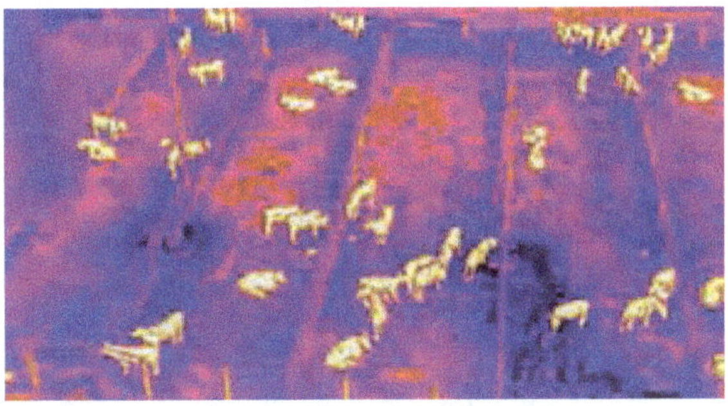

Figure 6. Heat map of herd of cattle [36].

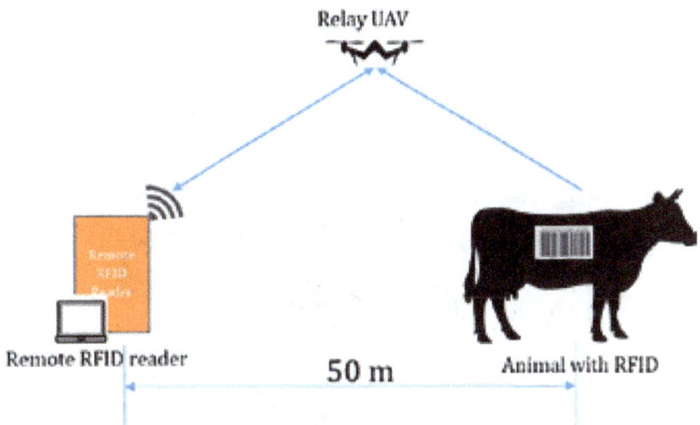

Figure 7. Animal identification using relay UAVs.

maneuverable and agile. The challenges with the use of helicopters are the need for extensive training, the cost of licenses and certifications, the cost of fuel and most especially the high level of risk exposure and casualties associated with it.

UAVs provide a unique opportunity for aerial mustering as they are comparatively risk free, cheaper to fly and require shorter training period, yet able to achieve similar results. UAVs have successfully been used in Australia and New Zealand to muster sheep and cattle [40]. According to [41], aerial mustering UAVs are fitted with sirens to herd sheep, deer and cattle. The UAVs can also be used to guide the animals to feeding, drinking and milking areas. Numerous case studies of the application of the DJI Phantom in Agriculture are given in [41]. **Figure 8** shows a use case of UAVs for sheep mustering.

4.5. Geo-fencing and virtual perimeter

Geo-fencing, virtual perimeter or geo-zoning simply means creating a virtual barrier or perimeter around a geographical area of interest [42, 43]. It has also been defined as an enclosure, or a boundary without the use of physical barriers. It can be accomplished by using a combination of RFID, LoRaWAN and GPS based location sensors for instance. Sensors obtain the location of the subject of interest relative to a map. Geo-fencing has been used in numerous fields such as fleet management and logistics – to monitor movement of vehicles, proximity marketing – which prompt users of products when they are close enough, asset management – which send alerts when an asset is moved without authorization, people monitoring – such as in monitoring movement of children and employees and in law enforcement – to restrict and/or monitor persons of interest.

Figure 8. Aerial mustering using a UAV.

Figure 9. A virtual fence around a grazing area.

Geo-fencing has also seen immense application in Agriculture, more specifically in free-range livestock farming. Sensors are placed on collars of cattle, goats etc. and these send location data to the farmer. There are two major forms of application of geo-fencing in agriculture: in the first, the sensors simply notify the farm owner when animals have grazed outside a pre-defined perimeter [44]. In this system, the farmer has to actively go muster the animals back into the perimeter. In the second approach, the animals are given subtle stimulations when they wonder outside set perimeter. Such simulations might include high frequency sounds or low voltage jolts – this approach depends on associative learning [45]. An illustration of a geo (virtual)-fence is shown in **Figure 9**, with the red boundary showing the grazing area and the blue circle showing an animal grazing outside the boundary.

Recent research work has focused on improving the efficiency of geo-fencing technologies. Low cost GPS being the most commonly used geo-fencing sensors have an error range of between 5 and 10 m and sometimes take long to locate and lock on to the required number of satellites. In a bid to improve on them, Assisted-GPS and WiFi have been used to respectively improve accuracies and reduce the time-to-first-fix [43]. LoRaWAN has recently been introduced as an alternative protocol for accurate location of animals [44, 46].

Though some arguments have been raised with respect to the effectiveness of geo-fencing, such as in the work of [47], rather than purely depending on stimuli, UAVs can be used to steer the animals back into range when they roam out of grazing perimeters. UAVs can therefore provide a cheap and effective way of getting animals back "inline" and are particularly useful when a number of animals stray outside different ends of the perimeter.

5. Advantages and challenges of UAVs

5.1. Advantages

1. Limited Constraints: Being air borne they are not hindered by physical constraints such as road/soil terrain, uneven paths and obstacles. They can simply fly over them all.

2. Shorter travel path: It is well known that the shortest distance between two points is a straight path. UAVs are best suited for this, as they can fly directly in straight paths. This is not always the case with land based vehicles.

3. Flying dark: In the case of autonomous UAVs, the UAVs can be programmed to fly in pitch darkness or at times with near zero visibility when it would be difficult for humans to manually control them.

4. Time and labor savings: Activities such as head count, monitoring and mustering often require the employment of more hands to help out. These can be both labor intensive and time consuming. With the use of UAVs, the number of extra laborers required is significantly cut down, while simultaneously saving time. Similarly in crop farming, UAVs can spray crops about 40–60 times faster than human laborers can.

5. Cost: Beyond savings in time, cutting down on laborers directly translates to cost savings. Though, capable UAVs are not cheap and there is also the added cost incurred in form of electricity to recharge the batteries; the cost savings and advantages of UAVs still significantly outweighs the manual and labor intensive processes of traditional/crude agriculture.

6. Aerial photography and imaging: With the use of UAVs, farmers can quickly obtain aerial images of their entire farm or select areas of interest. This can be useful in determining when fruits start to sprout or when pests and weeds are choking out crops.

5.2. Hindrances and challenges

UAVs have seen a wide range of applications in a smart city, all of which contributes greatly to the development of any smart city. In [48] the authors pointed out some of the challenges associated with the use of UAVs. Though these works focused on smart city applications, a number of these challenges are also applicable to the Agricultural space. The challenges were broadly classified into business and technical, and include:

1. Cost: The technology is perceived as expensive as a result of the technical nature of UAVs. Deployment, integration and training can be very expensive [48]. Similarly, in [49], the authors took a project management perspective to deployment of UAV related projects and highlighted cost as a key element that needs to be considered. It was also noted that proper estimations need to be using various technique prior to undertaking any such project.

2. Licensing and regulation issues: This is still a gray area with respect to UAVs. Regulations are either none-existent or a loose adaptation of aviation laws, which do not perfectly fit in with UAVs. There is therefore the need to draw up legislation to regulate the new possibilities and application areas of UAVs. Countries such as the USA, the UK, Germany and Spain [31] are leading the way in this direction by drafting guidelines for the use of UAVs and areas over which they can be flown. Other countries of the world are however still some way behind.

3. Business Adoption: From a business perspective, it might not be out rightly easy to justify the adaptation of UAVs into Agriculture. Though one might argue that there might be cost savings in the long run, counter arguments can be put forward regarding the actual acquisition cost of the UAVs, insurance / replacement of crashed UAVs, purchase of high resolution cameras for imagery as well as the accompanying software solutions and other running costs. When all these are added, it makes it a hard case to sell to farmers and Agriculture business owners.
4. Technical Challenges: These come in the form of system integration - integration of the middleware services with the UAV, high performance systems for data analytics, Net-centric infrastructure which enable any member of a team to control the UAV and retrieve imagery and sensor information in real time and application of machine learning / computation intelligence to identify and retrieve useful insights from the large pool of data.
5. Ethics and privacy: Some feel that the use of UAVs for monitoring and surveillance would lead to the invasion of their privacy. A lack of standard operational and technological procedures needed for safe performance of the UAVs is a great challenge. There could be GPS-jamming and hacking because of the vulnerabilities in the command and control of UAV operations.

5.3. Other open challenges

1. Limited flight time: UAV flight time is largely dependent on battery capacity. In most UAVs, particularly the multi-rotor, batteries can often times only sustain a flight time of between 10 and 30 minutes, and can be less when flown during high wind speeds. For activities such as crop spraying UAVs are only effective on hills, small areas, and in areas where other equipment cannot easily reach, for longer distance/range they are less efficient and even more costly than larger ground-based crop spraying equipment. The same challenge can be seen in the area of NDVI imaging, where farmers obtain NDVI images to assess the plant health. Alternative solutions are airplanes and satellites. While UAVs are the most cost-effective for small areas, they are currently not competitive against planes and satellites for larger areas.
2. There is the need to improve battery technology and find a way of using batteries with bigger capacity yet small footprint in UAVs. The use of solar photo-voltaic cells to power UAVs, such as [50] or the hybrid fixed-wing might be promising direction to be explored.
3. Limited payload size: Due to the small size of most UAVs, they are unable to carry a lot at once. This therefore limits their applications to basic aerial photography and observation. Though there are large-size UAVs such as [51], these are still limited in terms of flight time which is even further shortened when they are fully loaded. This limitation is prominent in application of UAVs in crop dusting (spraying pest/weed controlling chemicals or fertilizer on crops). Large gas powered monocopters might be a potential solution to this challenge.

4. Autonomy of UAVs: The possibilities of UAVs in Agriculture are numerous. However, most are currently being manually operated by humans. This limits their applications to certain times of the day when there is clear visibility. Advancements in computational intelligence specifically in areas of navigation, obstacle avoidance automatic sensing and actuation (performing pre-programmed tasks) can further accelerate the acceptance and usage of UAVs in livestock Agriculture.

5. Data Processing: Recent research works have shown the importance of data and information in almost all areas of human endeavor. Agriculture is certainly no exception. The use of UAVs as data gathering tools is still very much in its infancy. There is the need to develop effective techniques for data acquisition, data muling and most importantly converting these data to useful information. For instance, by observing the movement and body temperature of cows, farmers might be able to detect possible health related issues early on before they become fatal.

6. Empowering Farmers: In an article titled "on Drone technology as a tool for improving agricultural productivity", in [2] the authors identified empowering farmers as a vital process in improving agriculture. They concluded that it's one thing to have the technology and have the ability to gather billions of data for analysis, however all these are of no use, if they cannot be properly integrated and applied into agricultural business processes; where it can bring the much needed improvement. This can only be done by empowering the farmers themselves – either through formal class room education or informal practical demonstrations.

7. Cost: The ideal UAV for agriculture applications is one that has a good balance of durability, long flight time, stability and optional ability to fly autonomously. Such a device would cost much more than an average farmer might be able to afford. Most especially for farmers in developing countries. For those in much developed countries, there might also be the challenge of justifying how the purchase of such expensive devices can directly translate to measureable profit. To this end Farmers are still largely depend on manual ways of carrying out their farm operations.

8. Safety: There are also safety concerns with the use of UAV in Agriculture. One such is the UAVs' inability to recognize and avoid other airborne aircrafts and objects within the same airspace. This could result in collisions. Though obstacle avoidance is not too far-fetched, incorporating such features into basic UAVs would further drive up the already expensive cost of the UAVs.

9. Availability: There is also the problem of manufacturing, and meeting up with the demands for UAVs by farmers. This is largely expected since the industry is still exploring and testing Agricultural use cases. Manufacturing is being done on a small scale and the fixed costs remain high. In [22], it was pointed out that despite the numerous potential advantages of thermal remote sensing has over the optical RS in crop and soil monitory, there are a number of practical difficulties in its use. These include but not limited to atmospheric attenuation and absorption, calibration, climate conditions, crop growth stages as well as complex soil and plant interaction that have thus far limited its use in the agricultural sector.

6. Conclusion

Unmanned Aerial Vehicles or UAVs are essentially flying robots. Though initially designed for military use, they have are now widely used in various areas, from recreational sports, fire-fighting, flight simulations / trainings to toys for children. In this chapter we presented an application of UAVs to commercial Agriculture. We presented four major types of UAVs, and though the multi-rotor UAV with its ability to hover on spot and take-off and landing vertical may seem well suited for agriculture, its limited flight time is a major limitation. The hybrid-fixed-wing-motor-rotor might be a better fit. A detailed insight into the applications of UAVs in crop production and livestock farming was also presented. A prominent requirement for most UAV application in Agriculture is an integrated camera, as it allows images to be taken. Images are used in weed identification and control, soil analysis, animal monitoring, animal head counts, geo-fencing, mustering among others. Like most machines, UAVs have the advantage of doing repetitive and monotonous works better and more efficiently when compared to humans. Some advantages of applying UAVs in Agriculture were presented, some of which include limited path constraints, time saving and reduction in manual labor. However, there are a number of challenges limiting UAVs, most prominent among which is cost. UAVs that are well suited for Agriculture use are expensive. Operation and maintenance also come at a cost. It is therefore often difficult to convince farmers and Agriculture related stakeholders to integrate UAVs into their business. Beyond cost, battery limitations, safety and legal related issues are still major hurdles that need to be scaled before UAVs can find a strong foothold in agriculture.

Acronyms

CNN	convolutional neural network
ESSW	Mearly season site-specific weed management
GIS	geographic information system
GPS	global positioning system
LAI	leaf area index
LoRaWAN	long range wide area network
LPS	livestock production systems
NDVI	normalized difference vegetation index
OBIA	object-based image analysis
QR	codequick response code
RFID	radio frequency identification

SIFT	scale-invariant feature transformation
UAV	unmanned aerial vehicle
VTOL	vertical take off and landing

Author details

Chika Yinka-Banjo* and Olasupo Ajayi

*Address all correspondence to: cyinkabanjo@unilag.edu.ng

University of Lagos, Lagos, Nigeria

References

[1] Vroegindeweij BA, van Wijk SW, van Henten E. Autonomous unmanned aerial vehicles for agricultural applications. In: Proceeding. International Conference of Agricultural Engineering (AgEng). Zurich; 2014. p. 8

[2] Sylvester G. E-Agriculture in Action: Drones for Agriculture. Bangkok: Published by Food and Agriculture Organization of the United Nations and International Telecommunication Union; 2018

[3] Chapman S, Merz T, Chan A, Jackway P, Hrabar S, Dreccer M, et al. Pheno-copter: A low-altitude, autonomous remote-sensing robotic helicopter for high-throughput field-based phenotyping. Agronomy. 2014;**4**(2):279-301

[4] Sugiura R, Noguchi N, Ishii K. Remote-sensing technology for vegetation monitoring using an unmanned helicopter. Biosystems Engineering. 2005;**90**(4):369-379

[5] Gago J, Douthe C, Coopman R, Gallego P, Ribas-Carbo M, Flexas J, et al. UAVs challenge to assess water stress for sustainable agriculture. Agricultural Water Management. 2015;**153**:9-19

[6] DJI. DJI Agras MG-1P Series. [Video file]. 2019. Available from: www.dji.com/mg 1p/infor#specs

[7] AgEagle Aeriel Systems Inc. AgEagle RX-60 Taking Agriculture Intelligence to the Next Level. 2018. Available from: https://docs.wixstatic.com/ugd/89e3c5_e3de865b41b644fbb68adea13706723c.pdf?index=true

[8] AlphaUnmmanedSystems. Alpha 800 UAV Helicopter. 2018. Available from: https://alphaunmannedsystems.com/alpha-800-uav/

[9] ArcturusUAV. Jump 20. 2017. Available from: https://arcturus-uav.com/product/jump-20

[10] Tjahjowidodo T, Lee S. Tendon-sheath mechanisms in flexible membrane wing mini-UAVs: Control and performance. International Journal of Aerospace Engineering. 2017; **2017**:18

[11] Mazur M, PWC. Six Ways UAVs Are Revolutionalizing Agriculture. 2016. Available from: https://www.technologyreview.com/s/601935/six-ways-UAVs-are-revolutionizing-Agriculture/ [Accessed: 09 July 2019]

[12] Peña JM, Torres-Sánchez J, de Castro AI, Kelly M, López-Granados F. Weed mapping in early-season maize fields using object-based analysis of unmanned aerial vehicle (UAV) images. PLoS One. 2013;**8**(10):e77151

[13] Torres-Sanchez J, Lopez-Granados F, De Castro A, Pena-Barragan J. Configuration and specifications of an unmanned aerial vehicle (UAV) for early site specific weed management. PLoS One. 2013;**8**(3):e58210. DOI: 10.1371/journal.pone.0058210

[14] Shafian S, Rajan N, Schnell R, Bagavathiannan M, Valasek J, Shi Y, et al. Unmanned aerial systems-based remote sensing for monitoring sorghum growth and development. PLoS One. 2018;**13**(5):e0196605. DOI: 10.1371/journal.pone.0196605

[15] Zhang J, Basso B, Price RF, Putman G, Shuai G. Estimating plant distance in maize using unmanned aerial vehicle (UAV). PLoS One. 2018;**13**(4):e0195223. DOI: 10.1371/journal.pone.0195223

[16] Lv M, Xiao S, Tang Y, He Y. Influence of UAV flight speed on droplet deposition characteristics with the application of infrared thermal imaging. International Journal of Agricultural and Biological Engineering. 2019;**12**(3):10-17

[17] Yallappa D, Veerngouda M, Maski D, Palled V, Bheemanna M. Development and evaluation of drone mounted sprayer for pesticide applications to crops. IEEE Global Humanitarian Technology Conference. 2017. pp. 1-7. DOI: 10.1109/GHTC.2017.8239330

[18] Hentschke M, Freitas E, Hennig C, Veiga C. Evaluation of altitude sensors for a crop spraying drone. Drones 2, 3. MDPI. 2018. p. 25. DOI: 10.3390/drones2030025

[19] Xiongkui H, Bonds J, Herbst A, Langenakens J. Recent development for unmanned aerial vehicle for plant protection in East Asia. International Journal of Agricultural and Biological Engineering. 2017:**10**(3);18-30

[20] Guo T, Kujirai T, Watanabe T. Mapping crop status from an unmanned aerial vehicle for precision agriculture applications. The International Archives of the Photogrammetry, Remote Sensing and Spatial Information Sciences. 2012;**39**:B1

[21] Baluja J, Diago M, Balda P, Zorer R, Meggio F, Morales F, et al. Assessment of vineyard water status variability by thermal and multispectral imagery using an unmanned aerial vehicle (UAV). Irrigation Science. 2012;**30**:511-522. DOI: 10.1007/s00271-012-0382-9

[22] Khanal S, Fulton J, Shearer S. An overview of current and potential applications of thermal remote sensing in precision agriculture. Computers and Electronics in Agriculture. 2016;**139**(2017):22-32

[23] Gonzalez-Dugo V, Zarco-Tejad P, Nicolas E, Nortes P, Alarco J, Intrigliolo D, et al. Using high resolution UAV thermal imagery to assess the variability in the water status of five fruit tree species within a commercial orchard. Precision Agriculture. 2013. DOI: 10.1007/s11119-013-9322-9

[24] Piug E, Gonzalez F, Hamilton G, Grundy P. Assessment of crop insect damage using unmanned aerial systems: A machine learning approach. In: 21st International Congress on Modelling and Simulation; Gold Coast, Australia; 29 Nov–4 Dec 2015

[25] Jianwei Y, Tianjie L, Changchun L, Jiangqun Z. The application of unmanned aerial vehicle remote sensing in quickly monitoring crop pests. Intelligent Automation and Soft Computing. 2012;**18**(8):1043-1052

[26] Yinka-Banjo CO, Owolabi WA, Akala AO. Birds control in farmland using swarm of UAVs: A behavioural model approach. In: Science and Information Conference. Cham: Springer; 2018. pp. 333-345

[27] Steinfeld H, Mäki-Hokkonen J. A classification of livestock production systems. World Animal Review. 1995:83-94

[28] Seré C, Steinfeld H, Groenewold J. World Livestock Production Systems. Food and Agriculture Organization of the United Nations. FAO Publishing; 1996

[29] Yitbarek M, Berhane G. Livestock production systems and analysis: Review. AIJCSR. 2014;**1**(2):16-51

[30] Steinfeld H, Wassenaar T, Jutzi S. Livestock production systems in developing countries: Status, drivers trends. Revue Scientifique et Technique. 2006;**25**(2):505-516

[31] Chamoso P, González-Briones A, Rivas A, Bueno FDM, Corchado J. The use of drones in Spain: Towards a platform for controlling UAVs in urban environments. Sensors. 2018;**18**:1416

[32] Chamoso P, Raveane W, Parra V, González A. UAVs applied to the counting and monitoring of animals. Ambient Intelligence-Software and Applications. Cham, Switzerland: Springer; 2014. pp. 71-80

[33] Vayssade J, Arquet R, Bonneau M. Automatic activity tracking of goats using drone camera. Computers and Electronics in Agriculture. Elsevier; 2019;**162**:767-772

[34] Havens K, Sharp E. Thermal Imaging Techniques to Survey and Monitor Animals in the Wild: A Methodology. Academic Press; 2015

[35] Elias AR, Golubovic N, Krintz C, Wolski R. Where's the bear?-Automating wildlife image processing using IoT and edge cloud systems. In: 2017 IEEE/ACM Second International Conference on Internet-of-Things Design and Implementation (IoTDI); IEEE; 2017. pp. 247-258

[36] Texas A&M AgriLife. Drones could apply thermal imaging to identify sick livestock in feedlots. 2019. Available from: https://research.tamu.edu/2019/03/07/drones-could-apply-thermal-imaging-to-identify-sick-livestock-in-feedlots/ [Accessed: 01 June 2019]

[37] Webb P, Mehlhorn SA, Smartt P. Developing protocols for using a UAV to monitor herd health. In: Proceedings of the 2017 ASABE Annual International Meeting; Spokane, WA, USA; 16-19 July 2017

[38] Nyamuryekunge S, Cibils A, Estell R, Gonzalez A. Use of an unmanned aerial vehicle—Mounted video camera to assess feeding behavior of raramuri criollo cows. Rangeland Ecology & Management. 2016;**2016**(69):386-389

[39] Ma Y, Selby N, Adib F. Drone relays for battery-free networks. In: Proceedings of the Conference of the ACM Special Interest Group on Data Communication; ACM; 2017. pp. 335-347

[40] Man and Drone. Drone sheep heading in New Zealand [video file]. 2018. Available from: https://youtu.be/D8mXL2JapWM

[41] DJI Ferntech. Drones on the farm [Video file]. 2018. Available from: https://www.djistore.co.nz/agriculture

[42] Abaqus. GeoFencing and Alerts for Location Based Services. 2014. Available from: www.myGeoTracking.com. [Accessed: 21 July 2019]

[43] Rahate S, Shaikh M. Geo-fencing infrastructure: Location based service. International Research Journal of Engineering and Technology. 2016;**3**(11):1095-1098

[44] Andonovic I, Michie C, Cousin P, Janati A, Pham C, Diop M. Precision livestock farming technologies. In: 2018 Global Internet of Things Summit (GIoTS); IEEE; 2018. pp. 1-6

[45] Lee C, Henshall JM, Wark TJ, Crossman CC, Reed MT, Brewer HG, et al. Associative learning by cattle to enable effective and ethical virtual fences. Applied Animal Behaviour Science. 2009;**119**:15-22

[46] Waterhouse A, Duthie C, Kodam S. Overcoming challenges for geofencing of real-time monitored grazing livestock. In: Proc. BSAS Annual Conference, Dublin. Advances in Animal Bioscience, paper 219; 2018

[47] Lomax S, Colusso P, Clark C. Does virtual fencing work for grazing dairy cattle? Animals. MDPI. 2019;**9**(7):429. DOI: 10.3390/ani9070429

[48] Mohammed F, Idries A, Mohamed N, Al-Jaroodi J, Jawhar I. UAVs for smart cities: Opportunities and challenges. In: 2014 International Conference on Unmanned Aircraft Systems (ICUAS); IEEE; 2014. pp. 267-273

[49] Idries A, Mohamed N, Jawhar I, Mohammed F. Challenges of developing UAV applications: A project management view. In: Proceedings of the 2015 International Conference on Industrial Engineering and Operations Management; Dubai, United Arab Emirates (UAE); March 3-5, 2015

[50] Gibbs Y. Past Projects: Pathfinder/Pathfinder Plus Solar-Powered Aircraft. NASA. 2017. Available from: https://www.nasa.gov/centers/dryden/history/pastprojects/Erast/pathfinder.html [Accessed: 05 June 2019]

[51] Obrazcova T. Drones Give Air Cargo a New Buzz. Aviation Pulse. 2016. Available from: https://50skyshades.com/news/airlines/drones-give-air-cargo-a-new-buzz [Accessed: 05 June 2019]

3

Robust Outdoor Vehicle Visual Tracking Based on k-Sparse Stacked Denoising Auto-Encoder

Jing Xin, Xing Du, Yaqian Shi, Jian Zhang and Ding Liu

Abstract

Robust visual tracking for outdoor vehicle is still a challenging problem due to large object appearance variations caused by illumination variation, occlusion, and fast motion. In this chapter, k-sparse constraint is added to the encoder part of stacked auto-encoder network to learn more invariant feature of object appearance, and a stacked k-sparse-auto-encoder–based robust outdoor vehicle tracking method under particle filter inference is further proposed to solve the problem of appearance variance during the tracking. Firstly, a stacked denoising auto-encoder is pre-trained to learn the generic feature representation. Then, a k-sparse constraint is added to the stacked denoising auto-encoder, and the encoder of k-sparse stacked denoising auto-encoder is connected with a classification layer to construct a classification neural network. Finally, confidence of each particle is computed by the classification neural network and is used for online tracking under particle filter framework. Comprehensive tracking experiments are conducted on a challenging single-object tracking benchmark. Experimental results show that our tracker outperforms most state-of-the-art trackers.

Keywords: visual tracking, k-sparse stacked denoising auto encoder, classification neural network, robust visual tracking, particle filter

1. Introduction

The purpose of the visual tracking for outdoor vehicle is to estimate the state of outdoor vehicle and provide current traffic state accurately and comprehensively. At present, it has become an important part of intelligent transport system (ITS). However, robust tracking for outdoor vehicle is still a challenging problem due to the complex and varying outdoor environment. Many researchers proposed solutions to the different challenging environment. Rad [1] proposed a strategy that can solve the problem of occlusion during the tracking process of moving vehicles on highway. But, the tracking accuracy of this method will be greatly reduced when the lighting conditions change sharply. Zhang et al. [2] proposed a multi-layer occlusion detection and processing framework that can be used to deal with the problem of mutual occlusion between two vehicles. Faro et al. [3] further improved [2] by introducing curvature scale space to segment occluded region accurately. Xin et al. [4] proposed adaptive multiple cues integration for robust outdoor vehicle visual tracking in the particle filter framework. This method has strong robustness against color interference and partial or complete occlusion of vehicles. Although these existing methods have achieved certain progress in outdoor vehicles visual tracking, these methods can only deal with the occlusion problem between two vehicles or the occlusion of vehicles by other objects. However, in the actual traffic scene, the mutual occlusion between multiple vehicles often occurs and faces the challenges of complex outdoor environments such as illumination variation (IV), cluttered background (BC), and fast motion (FM). Therefore, the robust outdoor vehicles visual tracking remains a thorny issue.

Existing visual object tracking algorithms are mainly divided into two major categories that include generative model and discriminative model. The generative model learns the appearance representation of the object and searches the candidate area that most closely matches the object appearance template as the location of the object in the new frame. The discriminative model treats the object tracking as a binary classification problem, using the learned characteristics to distinguish the object and background information. Therefore, the extraction of robust features is the key to the success of the object tracking technology. Traditional visual object tracking methods rely on artificial features; the low-dimensional artificial features are not robust to large appearance variation of object. Recently, deep learning shows promising performance in automatic extracting feature that outperforms pre-defined handcraft feature methods. Deep learning can map the original feature space to another feature space to learn more abundant features. Recently, deep learning has been widely applied to image processing, speech recognition, natural language processing, health care, robotics, and other fields for its powerful feature learning capability. It has been proved that feature representation when learnt in a deep learning way encourages sparsity. And k-sparse constraint can guarantee that each input for a certain sparsity. At the same time, some scholars have applied it to video object tracking technology. Due to the powerful feature representation ability of deep learning, the robustness of visual object tracking technology has been greatly improved. Wang et al. [5] proposed a fully convolutional networks tracker (FCNT) that uses convolutional neural networks to learn the characteristics of objects from large-scale classification datasets and further analyses performance of the extracted features in the object tracking aspect. High-level features are good at distinguishing different kinds of objects and are very robust to the appearance variation of the object. Low-level features more focus on the local details of the object and can be used to distinguish similar distractors in the background. FCNT can effectively prevent object tracking drift based on the effective use of different layers of convolutional neural network (CNN) features. Nam et al. [6] subsequently proposed the Multi-Domain Convolutional Neural Networks (MDNet). Unlike [5], MDNet directly uses the tracking video data sets train, the deep learning model to obtain the universal feature representation of the object, and then fine-tunes the network parameters for each particular video sequence in online tracking to achieve more robust tracking. However, the tracking speed of MDNet is slow and cannot meet the requirements of real-time performance. Existing research has demonstrated that sparseness is encouraged when deep learning learns feature representations. Because sparse representation can reduce the complexity of the representation, which is crucial to improve the speed of the object tracking algorithm, sparse constraints can be used to further optimize the deep network [7, 8] and can make the original signal express more meaningful, which has been verified by independent principal component analysis and sparse coding algorithm [9]. In general, there are two ways to add sparse constraints into the deep network for sparse representation: sparseness of the hidden layer response and weight sparseness between the hidden layer and the input layer. In this chapter, we adopt the first method for sparse representation. At the same time, we perform k-sparse constraint in neural network to keep only k highest activities in hidden layers, which can maintain the sparse representation of each input [10]. In other words, we add the k sparse constraint to the original stacked denoising auto-encoder (SDAE) hidden layer unit and form kSSDAE, which is used as a feature extractor in the target tracking to better learn the invariant feature of the object appearance. Therefore, the application of kSSDAE in object tracking can overcome poor robustness problem and further improve the robustness of visual tracking. The main contributions of this chapter are as follows.

- We propose a new auto-encoder–based tracking method, namely kSSDAE tracker, to solve the robust tracking for outdoor vehicles in complex environments, such as occlusion, clutter background, illumination variation, and so on.
- We add the k-sparse constraint into the encoder part of stacked auto-encoder network to learn more invariant feature of object appearance during the tracking.
- We evaluate our method on a challenging single-object tracking benchmark with 51 video sequences and 11 attributes.

2. Related works

Deep learning has exhibited powerful automatic feature extraction capability in computer vision tasks such as image classification, object detection, and so on. Visual object tracking is one of the important research contents in the field of computer vision. The performance of the tracker can be greatly improved due to the applications of the deep learning. Currently, two kinds of deep learning models including convolution neural network and deep auto-encoder are mainly used in the visual object tracking to perform automatic feature extraction.

2.1. Convolutional neural network for object tracking

Convolutional neural network (CNN) is a multi-layered supervised learning feedforward neural network. A typical CNN structure includes convolutional layer, pooling layer, and full connection layer. Specially, the automatic feature extraction function of the CNN is mainly realized through the convolution layer and pooling layer. The structure of the CNN determines that it has natural advantages for image processing, and it also shows a competitive performance in visual tracking. In order to solve the problem of object drift caused by similar or clutter background in visual tracking, Fan [11] et al. use CNN to learn spatial and temporal invariance features between adjacent frames. Jin [12] combine a CNN with two convolutional layers and two pooling layers and radial basis function (RBF) to perform feature extraction so that it can better learn the invariable features of the object appearance in visual tracking. Hong [13] use an offline-trained CNN to extract the distinctive feature map of the object in visual tracking. Wang [14] train a two-level CNN by offline way and use it for online object tracking. The network pays more attention to the learning of motion invariant features. Unlike most CNN used for object tracking, the network designed by Wang [15] et al. is not a binarized output classification result but instead generates a probability map to represent the potential area of the object. The use of CNN greatly improves the accuracy of visual tracking, but high computational complexity is still a limitation. In order to improve the real-time performance of the tracking algorithm, Doulamis et al. [16] proposed a fast adaptive supervised algorithm for object tracking and classification. In addition, although the pooling operation in CNN can obtain invariant features to drop the recognition effect caused by the change of the object appearance, however, it reduces the resolution of the image and leads to spatial information loss. The loss of information of pooling operations is crucial for tracking [17]. Zhang et al. [18] combined convolutional neural networks with spatial-temporal saliency-guided sampling for object tracking in a correlated filter framework. The algorithm establishes an optimization function to locate object positions based on significant region detection and significant motion estimation. Different from other object tracking algorithms whose location estimation is based on the last layer of the convolutional neural network, this algorithm combines intra-frame appearance correlation information and inter-frames motion saliency information to ensure accurate target location. All in all, the object tracking algorithm based on the convolutional neural network can effectively track object, but the network structure is relatively complex, consumes a lot of training time, and requires a large number of labeled training samples, and it is difficult to achieve a balance between tracking accuracy and tracking speed.

2.2. Deep auto-encoder for object tracking

The basic idea of the deep auto-encoder (DAE) is to encode the input signal and then use the decoder to reconstruct the original signal. The goal of the one is to minimize the reconstruction error between the reconstructed signal and original signal. Compared with the method of visual tracking using CNN, a DAE compresses the original signal by coding, removes redundancy, and can reflect the more primitive nature of the original signal in a more concise manner. Therefore, visual tracking using DAE has a lower calculation cost and is more suitable for some occasions with high real-time requirements. In 2013, Wang et al. [19] proposed a novel deep learning tracker (DLT), which for the first time uses a DAE for tracking. DLT considers the object tracking task as a two-category problem. Firstly, using Tiny Images data set to offline train a stacked denoising auto-encoder (SDAE) in an unsupervised manner to obtain a universal image feature representation for object and then use it for online tracking. The classification neural network is constructed and is fine-tuned in the tracking process to distinguish the target from the background. Soon after, many improved versions of the DLT methods have been proposed. For example, Zhou [20] combined online AdaBoost feature selection framework with SDAE for object tracking to effectively solve complex and dramatic changes of the object appearance. Cheng et al. [21] used the SDAE network to implement adaptive target tracking in an incremental deep learning approach under the dual particle filter framework. Cheng et al. [22] implemented an object tracking algorithm based on enhanced group tracker and SDAE in the framework of the popular tracking-learning-detection (TLD) algorithm in order to solve the object drift of the tracking method based on the appearance model. Due to the Haar-like features in the multi-instance learning (MIL) tracking algorithm are difficult to reflect the shortcomings of the object itself and the external changes, Cheng et al. [23] introduced SDAE to extract the effective features of the example image to achieve higher precision tracking. In order to further improve the application performance of the stacked denoising auto-encoder in video object tracking, some scholars have proposed many improved tracking algorithms based on a stacked denoising auto-encoder. Dai et al. [24] proposed a local patch tracking algorithm based on a stacked denoising auto-encoder. The algorithm partitions the input image; then a feature extractor combining multiple stacked denoising auto-encoder is used to describe the feature information of local patch and fuse their local features to achieve object tracking. The local feature extraction greatly reduces the computation complexity compared with the global feature representation. In the tracking process, the weight of each patch of the object candidate region can be adaptively adjusted according to the confidence of the corresponding network. Hua et al. [25] proposed a new visual tracking algorithm based on the multi-level feature learning capability of the stack denoising auto-encoder under the particle filter framework.

The training of the stacked auto-encoder network includes two stages of hierarchical pre-training and online tracking. In the hierarchical pre-training stage, a description of multi-level image features is obtained. In the online tracking stage, the network parameters are back-propagated through the genetic algorithm to fine-tuning. The use of genetic algorithm in network parameter adjustment effectively avoids the deficiency of traditional BP algorithm and further enhances the robust performance of the network. These trackers can use SDAE for unsupervised feature learning on data that lacks tagging, improving the problem of insufficient training data for deep neural networks (DNNs). However, in some challenging and complex environments, these trackers will fail to track the object. Therefore, we can further enhance the feature expression capabilities of deep neural networks (DNNs) for more robust tracking.

In this chapter, we add the K-sparse constraint into the coding part of the SDAE to learn more invariant feature of object appearance and propose a staked k-sparse-auto-encoder–based robust tracking algorithm for outdoor vehicle under particle filter framework to solve the problem of large appearance variations during the tracking.

3. The kSSDAE-based tracker

Overall structure of the proposed kSSDAE-based robust tracking algorithm for outdoor vehicle is shown in **Figure 1**. The tracking system mainly includes three parts as follows: offline

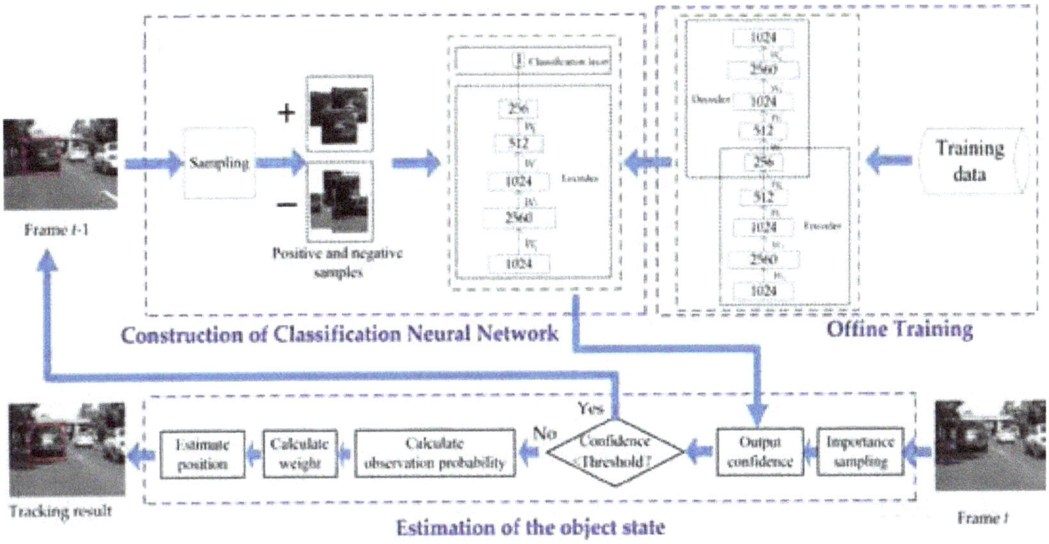

Figure 1. Overall structure of the proposed kSSDAE-based robust tracking algorithm for outdoor vehicle.

training of SDAE, construction of classification neural network, and estimation of object state. The basic idea of the algorithm is: firstly, we adopt the pre-trained SDAE model proposed in DLT [15] to learn the generic feature representation. Training data of the model are obtained through sampling randomly 1 million images from Tiny Images data set [26]. Tiny Images data set contains many kinds of the scene image. Before offline training, we need to pre-process the input data with 32×32. Offline training way of the SDAE is unsupervised. Secondly, we propose a kSSDAE model to learning more invariant feature of object appearance during tracking and train a classification neural network to compute the confidence of each particle. This is the key step to achieve robust tracking. Without the kSSDAE, the input cannot be guaranteed to have a sparse representation to extract more effective features to adapt the object appearance change. Finally, we estimate the object state under the particle filter framework, that is, the object state of the current frame can be represented by the particle with maximum confidence, which is calculated by classification neural network.

The specific implementation of the two main parts of the proposed tracking method will be stated in detail in the next section.

3.1. Construction of classification neural network

The main function of this module is to compute the confidence of each particle during the online tracking. Here, confidence is used for evaluating every particle's reliability. In this chapter, the classification neural network can be constructed by connecting the encoder of the well-trained kSSDAE with a classification layer as shown in **Figure 2**.

In feedforward phase, the hidden activities function z can be computed as

$$z = W^T x + b \tag{1}$$

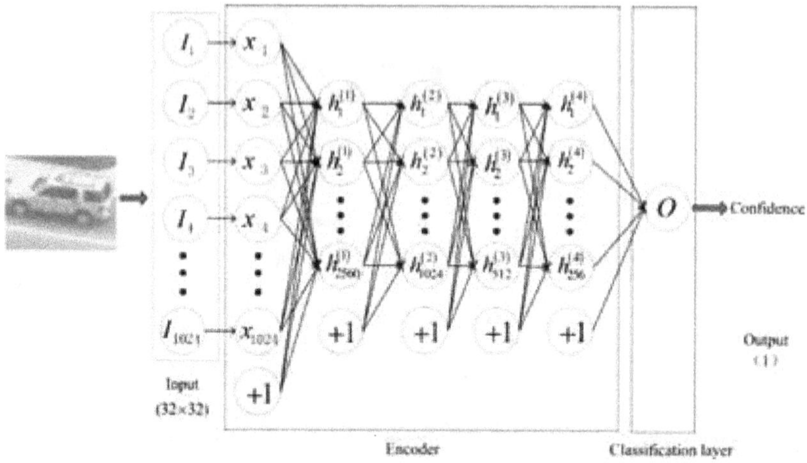

Figure 2. Architecture of classification neural network (1024-2560-1024-512-256-1).

where, x is the input vector, W is weight, and b is bias. We keep the k as largest hidden units and set others to zero.

Reconstruction error can be computed using the sparsified z as follows:

$$E = \left\| x - \left(Wz + b' \right) \right\|_2^2 \tag{2}$$

In back propagate phase, weights can adjusted by the k highest activities back propagating the reconstruction. The confidence computed by classification neural network reflects the credibility of decision in feature vector space of classifier. Ref. [27] has proved that when we use mean square error or cross-entropy as the cost function, the output expectation of multi-layer neural network is posterior probability of each class.

Let o_i be the output of the neural network corresponding to the k_i class, the output expectation can be computed by the posterior probability

$$E(o_i) = P(k_i|x) \tag{3}$$

Generally, the class with maximum probability is taken as a decision. So, the confidence can be obtained from the maximum output of the classification neural network.

$$c(x) = E(\max o_i) \tag{4}$$

At the beginning of the visual tracking, we select the object to be tracked and fine-tune the classification neural network using positive and negative samples. In the process of online tracking, in order to adapt specific object appearance changes, we need to fine-tune the classification neural network again when the confidence, which is calculated by the classification neural network, is lower than the predefined threshold.

3.2. Estimation of the object state

Object state can be estimated by the object tracking algorithm, which can be viewed as a problem to estimate the posterior distribution $p(s_t^i|y_{1:t})$ of state s_t at time t according to dynamic system $p(s_t^i|s_{t-1})$ of the object state. In this chapter, object state s_t is represented by six affine transformation parameters corresponding to horizontal translation, vertical translation, scale angle, aspect ratio, and skewness, and the state transition distribution $p(s_t^i|s_{t-1}^i)$ of each dimension can be modeled as a zero-mean normal distribution. The purpose of visual object tracking is to estimate the object state s_t (location, scale, etc.) from image sequences given all observations by any appropriate loss function, for example, maximum a posteriori (MAP) estimation or minimum mean square error (MMSE) estimation. The main online tracking steps under the particle filter framework are as follows.

3.2.1. Computing observation probability

Each particle represents a possible instantiation of the state of the object being tracked. Most likely, particle represents the object state at time t. Confidence c_t^i of each particle can be calculated by the classification neural network. When the maximum confidence is lower than the predefined threshold τ, that is, if $\max(c_t^i) \leq \tau$, we will fine-tune classification neural network by reselecting positive and negative training samples. If $\max(c_t^i) > \tau$, we calculate the observation probability by normalizing confidence

$$p(y_t|s_t^i) \propto c_t^i, i = 1, 2, \ldots, n \tag{5}$$

3.2.2. Updating weight

The weights for each particle can be updated according to the observation probability

$$w_t^i = w_{t-1}^i \cdot \frac{p(y_t|s_t^i)p(s_t^i|s_{t-1}^i)}{q(s_t|s_{t-1}, y_{1:t})} \tag{6}$$

where, $q(s_t|s_{t-1}, y_{1:t})$ is importance distribution and is often assumed to follow a first-order Markov process in which the state transition is independent of the observation. So the weights are updated as $w_t^i = w_{t-1}^i \cdot p(y_t|s_t^i)$.

Finally, object state can be estimated by taking the particle with the largest weight at each time step.

The implementation process of the proposed kSSDAE-based tracker is given as follows:

Algorithm Outdoor Vehicle Tracking

Input: Training samples; Video frame t.

Training SDAE offline;

Constructing classification neural network;

Connecting the encoder part of kSSDAE and a classification layer as shown in **Figure 2**;

Adding k sparse constraint into classification neural network;

For $t = 1, 2, \ldots, N$ frame number **do**

Sampling particles $S_t = \{s_t^i\}_{i=1}^{n}$;

Calculate confidence for each particle by (4);

If $t = 1$

Sampling positive and negative samples;

Fine-tuning classification neural network;

end

If $\max(c_t^i) \leq \tau$

Sampling positive and negative samples;

Fine-tuning classification neural network.

Else

Calculating observation probability by (5);

Updating weights by (6);

$t = t + 1$.

end

end

Output: Object state

4. Experiments

In this section, we conducted a quantitative experiment to evaluate the proposed tracker (kSSDAE-T) on a popular single-object online tracking benchmark [28]. The benchmark data set provides 51 fully annotated video sequences that have the 11 challenging attributes. Most of these attributes exist in the real scene of outdoor vehicles. In order to better demonstrate the performance of our tracker, we compare our tracker with other three popular trackers, including deep learning tracker (DLT) [15], multi-task tracker (MTT) [29], and Circulant Structure of Tracking-by-Detection with Kernels (CSK) [30].

The main related parameters in our experiment are set as follows.

- Learning rate is set to 0.2; sparsity k is set to 40.

- Standard deviation of the conservation likelihood σ is set to 0.001.
- The number of particles is 1000, and the particle's confidence threshold τ is set to 0.8.
- We use momentum gradient method to optimize the network parameters, and the momentum parameter is set to 0.5.

4.1. Quantitative evaluation

In this chapter, we adopt two quantitative evaluation indicators: one-pass evaluation (OPE) of tracking precision and success rate [28]. The precision takes the position error as the benchmark, and the precision plot shows the percentage of frames whose estimation position error is less than the given threshold, and the horizontal axis of the precision plot is scaled to the range [0,50]. The success rate is based on the overlap rate, and the success plot counts the number of successful frames whose overlap is greater than the given threshold, and the horizontal axis of the success rate is scaled to the range [0,1]. We use the score for the threshold = 20 pixels of each precision plot and the area under curve (AUC) of each success plot to rank trackers and one-pass evaluation (OPE) for robustness evaluation. The scores and rankings of precision and success rate for four trackers on the overall performance and the 11 attributes performance are shown in **Table 1**, and the best tracking results corresponding to the overall performance and the 11 attributes are marked in bold, and the ranking score is shown after "\." In **Table 1**, SV: scale variation, OV: out-of-view, OPR: out-of-plane rotation, OCC: occlusion, LR: low resolution, IPR: in -plane rotation, IV: illumination variation, DEF: deformation, MB: motion blur, BC: background clutters, FM: fast motion. The precision plot and success rate plot of four trackers on overall performance is shown in **Figure 3**. The precision plots and success plots of

	kSSDAE-T (Ours)		DLT		CSK		MTT	
	Precision	Success rate	Precision	Success rate	Precision	Success rate	Precision	Success rate
Overall	**0.585\1**	**0.522\1**	0.550\2	0.499\2	0.545\3	0.443\4	0.475\4	0.445\3
SV	0.597\2	0.535\2	**0.602\1**	**0.547\1**	0.503\3	0.352\4	0.461\4	0.398\3
OV	**0.571\1**	0.537\2	0.526\2	**0.552\1**	0.379\3	0.410\3	0.374\4	0.392\4
OPR	**0.576\1**	**0.492\1**	0.527\3	0.464\2	0.540\2	0.439\3	0.473\4	0.423\4
OCC	**0.545\1**	**0.504\1**	0.532\2	0.502\2	0.500\3	0.404\4	0.426\4	0.422\3
LR	0.383\3	0.358\3	0.309\4	0.297\4	0.411\2	0.397\2	**0.510\1**	**0.506\1**
IPR	**0.551\1**	**0.479\1**	0.502\4	0.439\4	0.547\2	0.457\3	0.522\3	0.463\2
IV	**0.543\1**	**0.480\1**	0.514\2	0.472\2	0.481\3	0.388\3	0.351\4	0.337\4
DEF	**0.500\1**	**0.422\1**	0.433\3	0.389\2	0.476\2	0.370\3	0.332\4	0.334\4
MB	**0.359\1**	0.309\3	0.328\3	0.321\2	0.342\2	**0.336\1**	0.308\4	0.288\4
BC	0.528\2	0.440\2	0.455\3	0.398\4	**0.585\1**	**0.491\1**	0.424\4	0.411\3
FM	**0.460\1**	**0.421\1**	0.417\2	0.418\2	0.381\4	0.380\4	0.401\3	0.385\3

Table 1. Tracking performance on four trackers.

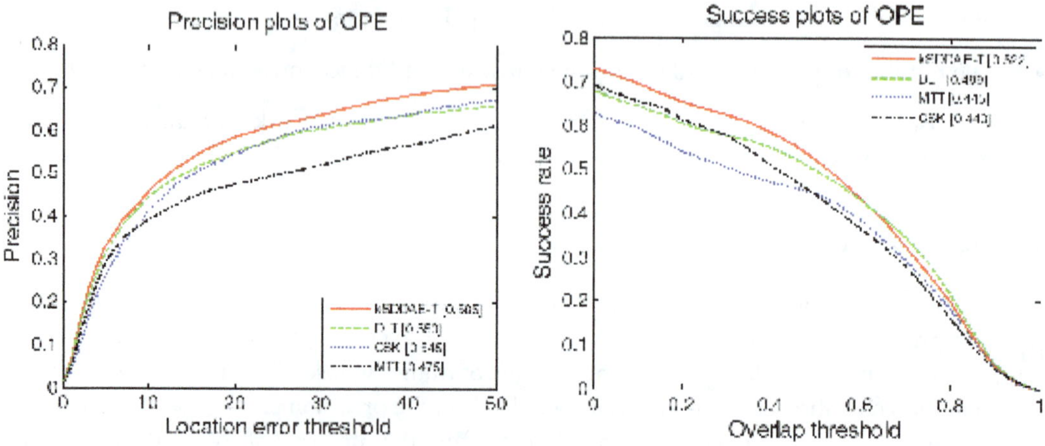

Figure 3. The overall performance of precision plot and success rate plot on four trackers.

four trackers on 11 attributes performance are shown in **Figure 4**. In order to analyze the performance of the tracker in every challenging attribute, [28] has marked the characteristics of each sequence and constructed subsets of the sequences with different saliency characteristics. For example, the OCC subset includes 29 sequences; it can be used for analyzing the ability of the tracker to handle occlusion problem. In **Figure 4**, the number that appears in the legend of each graph represents the ordinal number of sequence subset.

In overall performance of precision and success rate, our tracker is significant higher than the other three trackers. The performance of our tracker ranks first in 8 out of 11 attributes on precision as shown in **Table 1**. At the same time, the performance of our tracker ranks first in 6 out of 11 attributes on success rate. For the other attributes, except for LR attribute, our tracker has the success rate very close to the best on SV and BC attributes. The success rate of our tracker ranks 3 on MB attribute, but it is only lower than the best (CSK), 2.7%. Therefore, it can be concluded that the proposed kSSDAE-T tracker is the best compared with DLT, CSK, and MTT.

According to the precision plot and success rate plot of the four properties based on OCC, IV, MB, and FM attributes, we can see that our tracker can handle the appearance changes caused by most of outdoor environmental factors.

All in all, the proposed tracker can learn the invariable features of the object appearance and deal with the problem of object appearance changing caused by most of the outdoor complex environments. It can achieve better tracking results under most outdoor challenging conditions.

4.2. Qualitative evaluation

In order to further verify the effectiveness of the proposed tracking method in real scenarios, we compared the four trackers (proposed kSSDAE-based tracker, DLT, CSK, and MTT) on four outdoor vehicle sequences in real scenarios (Car4, CarDark, CarScale, and Suv). The attributes

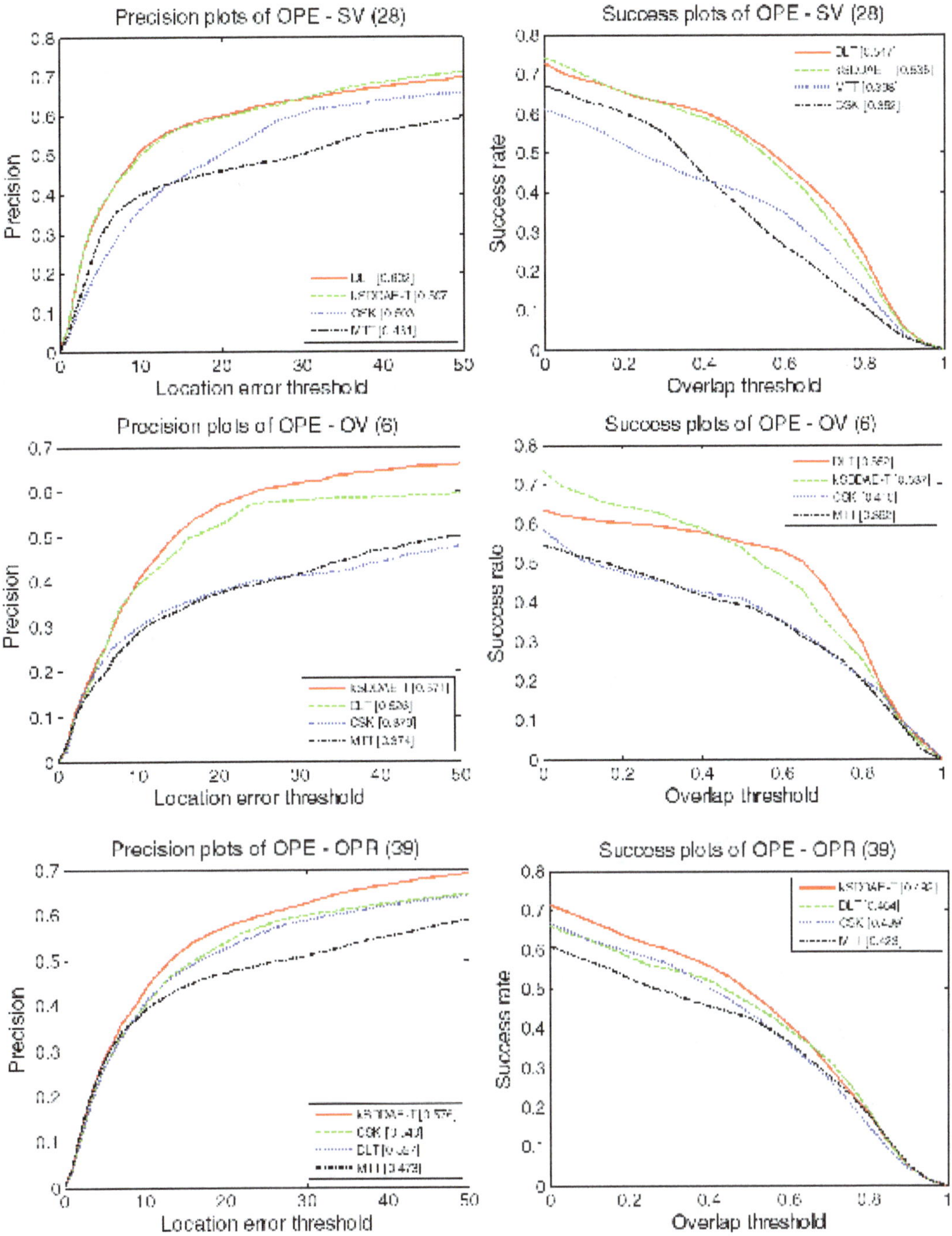
Precision plots of OPE - SV (28)
Precision
Location error threshold
Success plots of OPE - SV (28)
Success rate
Overlap threshold
Precision plots of OPE - OV (6)
Precision
Location error threshold
Success plots of OPE - OV (6)
Success rate
Overlap threshold
Precision plots of OPE - OPR (39)
Precision
Location error threshold
Success plots of OPE - OPR (39)
Success rate
Overlap threshold

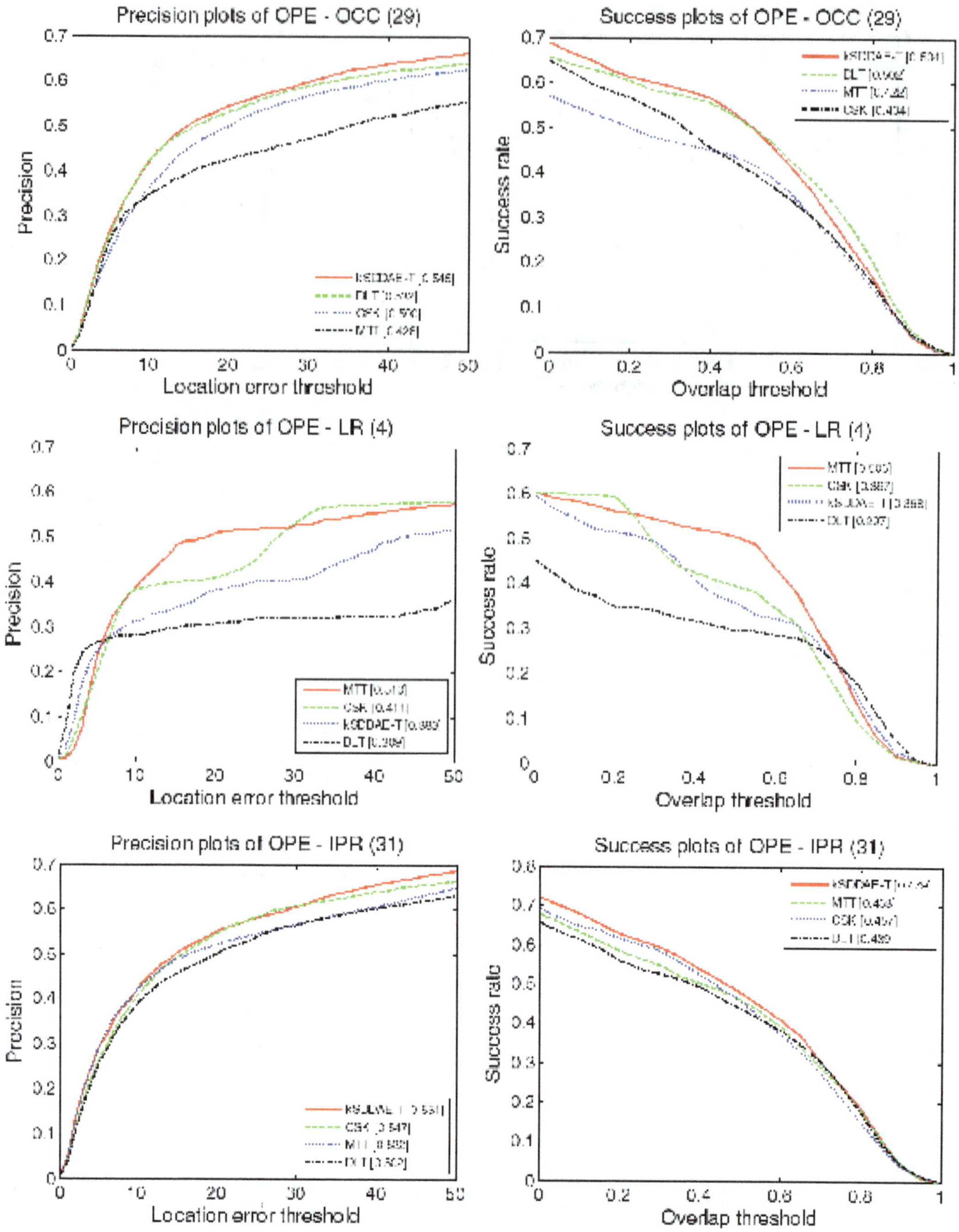
Precision plots of OPE - OCC (29)
Precision
Location error threshold
Success plots of OPE - OCC (29)
Success rate
Overlap threshold
Precision plots of OPE - LR (4)
Precision
Location error threshold
Success plots of OPE - LR (4)
Success rate
Overlap threshold
Precision plots of OPE - IPR (31)
Precision
Location error threshold
Success plots of OPE - IPR (31)
Success rate
Overlap threshold

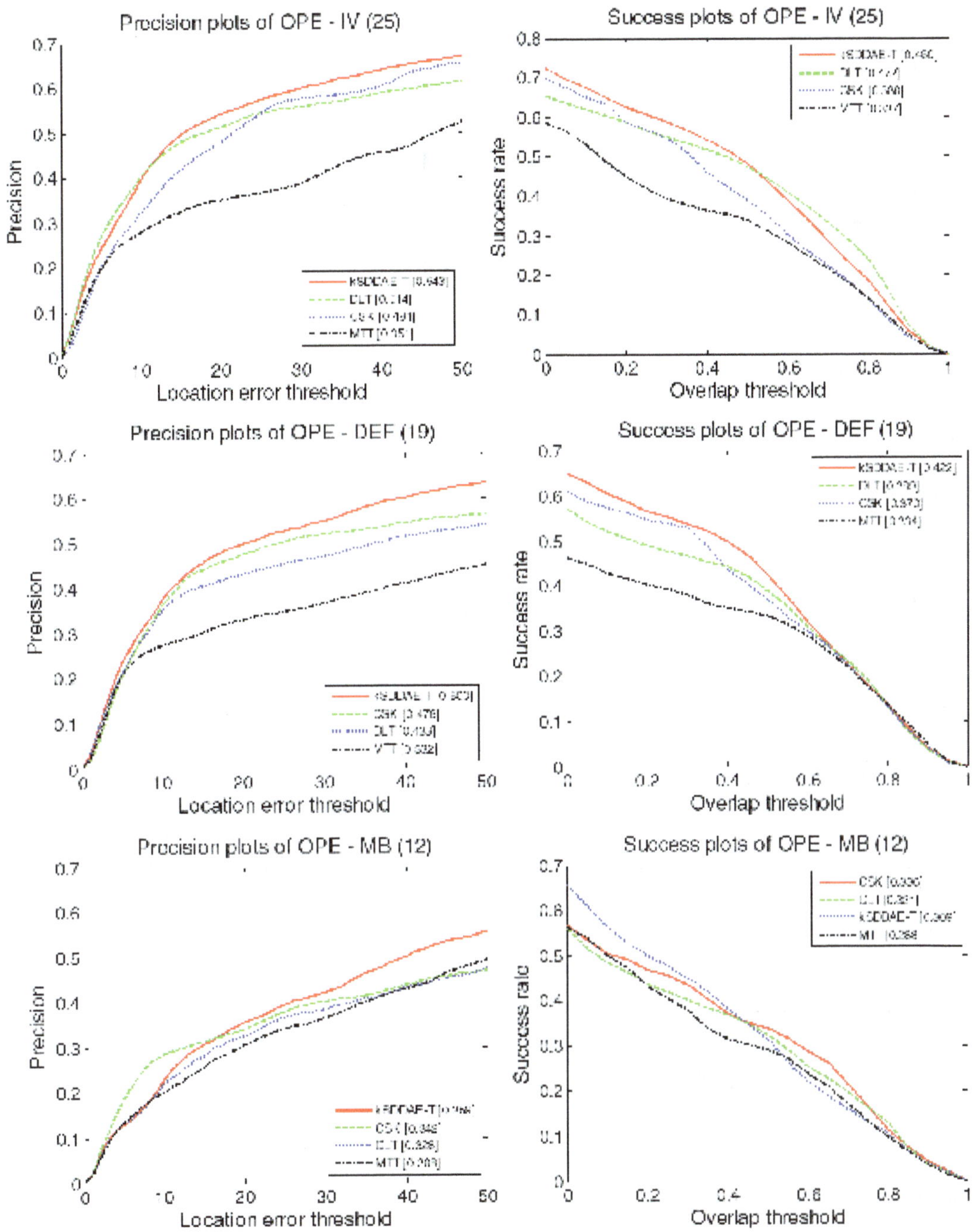
Precision plots of OPE - IV (25)
Precision
Location error threshold
Success plots of OPE - IV (25)
Success rate
Overlap threshold
Precision plots of OPE - DEF (19)
Precision
Location error threshold
Success plots of OPE - DEF (19)
Success rate
Overlap threshold
Precision plots of OPE - MB (12)
Precision
Location error threshold
Success plots of OPE - MB (12)
Success rate
Overlap threshold

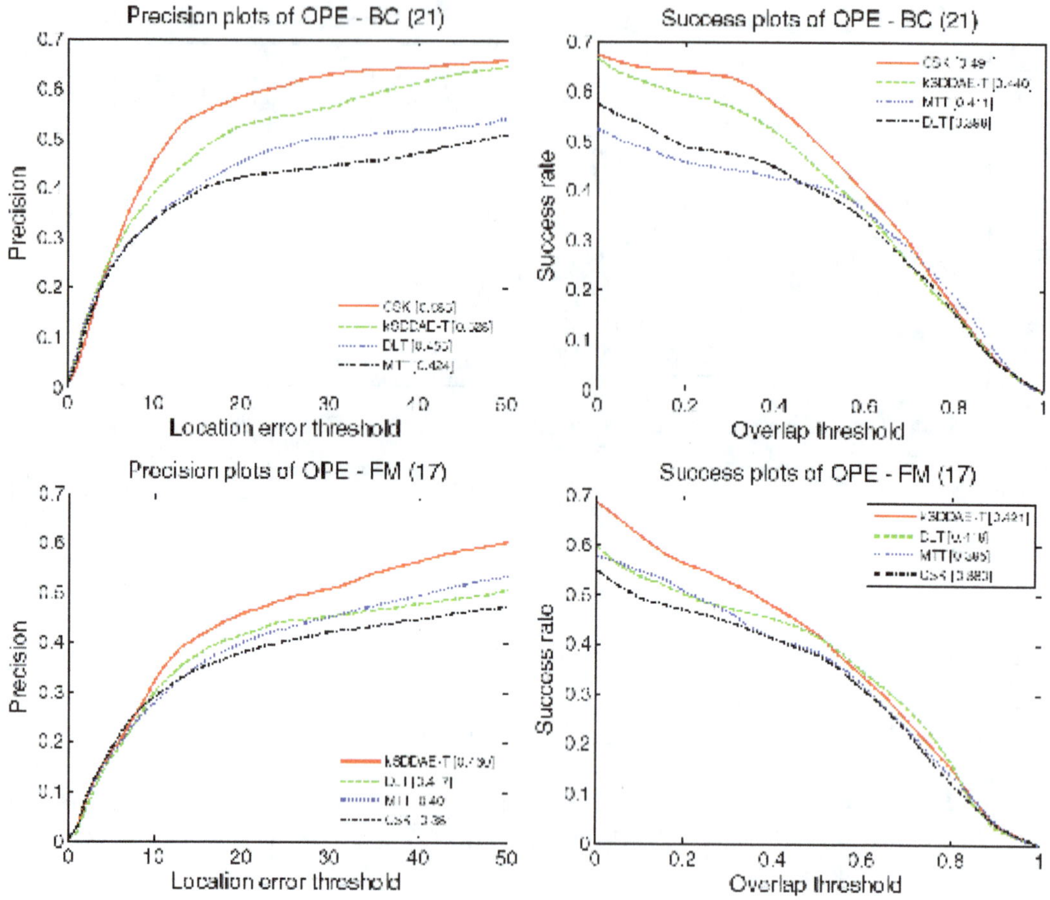

Figure 4. The precision plots and success plots of four trackers on 11 attributes performance (SV, OV, OPR, OCC, LR, IPR, IV, DEF, MB, BC, and FM).

Sequence	Attribute
Car4	IV, SV
CarDark	IV, BC
CarScale	SV, OCC, FM, IPR, OPR
Suv	OCC, IPR, OV

Table 2. Attributes of four sequences.

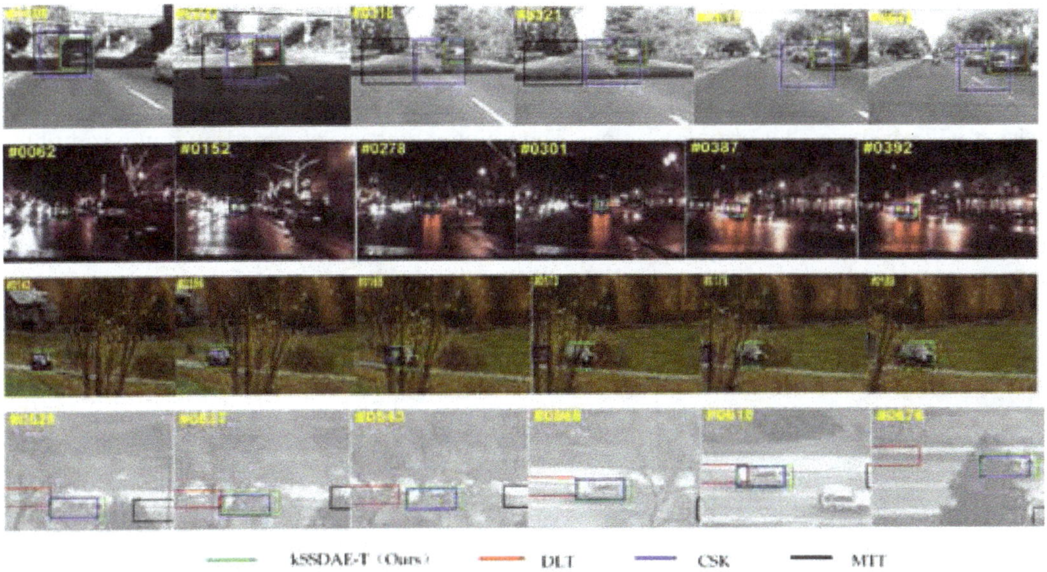

Figure 5. The sampled tracking results. Frame numbers are shown in the top left of each figure.

of four sequences are listed in **Table 2**. The partial tracking results of the four video sequences are shown in **Figure 5**.

In the video sequence Car4, when emerging IV and OCC near the #186, #233, and #318 frames, it can be seen from the tracking results that the CSK and the MTT tracker have different degrees of object drift. But, our tracker and DLT have achieved better tracking results. In addition, our tracker can also accurately track the target vehicle when SV emerges in #321, #612, and #635 frames. In the video sequence CarDark, our tracker can still perform effective tracking when the IV and BC emerging in #62, #152, #278, #301, #387, and #392 frames, while the MTT and CSK trackers have track drift at #301 frame, and at #387 frame, they completely lost the target vehicle. In the video sequence CarScale, our tracker can still show great performance when OCC was occurred in #165 and #175 frames, but CSK tracker failed. In the video sequence Suv, despite the OCC and similar background interference, our tracker can still accurately track the target vehicle.

To summarize, the proposed kSSDAE-based tracker can perform well in most complex outdoor environment.

5. Conclusion

In this chapter, we propose an improved auto-encoder–based approach for robust outdoor vehicle visual tracking. Our tracker can adapt the change of the object appearance during the tracking. The quantitative analysis on a standard evaluation platform shows that our tracker has a better tracking performance compared with the other three state-of-the-art trackers and has higher tracking precision in most of the outdoor vehicle tracking challenges. The qualitative analysis on four outdoor vehicle sequences in real scenarios shows that our tracker can work well in most complex outdoor environment.

The unsupervised training of kSSDAE requires that bottom image cannot be too large, otherwise it will consume a lot of training time. The training data are obtained by down-sampling directly from a full-sized image leading to information loss. In order to avoid loss of input image information, we can further improve the performance of outdoor vehicle tracking algorithms by using stacked convolutional auto-encoder (SCAE) [31] to take the outdoor vehicle tracking algorithm into application of life and industry.

Note: this chapter is an extended version of [32].

Acknowledgements

This work is supported by the Chinese National Natural Science Foundation under Grant No. 61873200,61833013,61573282.

Author details

Jing Xin[1]*, Xing Du[1], Yaqian Shi[1], Jian Zhang[2] and Ding Liu[1]

*Address all correspondence to: xinj@xaut.edu.cn

1 Shaanxi Key Laboratory of Complex System Control and Intelligent Information Processing, Xi'an University of Technology, Xi'an, P.R. China

2 Faculty of Engineering and Information Technology, University of Technology (UTS), Sydney, Australia

References

[1] Rad R et al. Real time classification and tracking of multiple vehicles in highways. Pattern Recognition Letters. 2005;**26**(10):1597-1607

[2] Zhang W, Wu QMJ, et al. Multilevel framework to detect and handle vehicle occlusion. IEEE Transactions on Intelligent Transportation Systems. 2008;**9**(1):161-174

[3] Faro A, Giordano D, et al. Adaptive background modeling integrated with luminosity sensors and occlusion processing for reliable vehicle detection. IEEE Transactions on Intelligent Transportation Systems. 2011;**12**(4):1398-1412

[4] Xin J, Liu XD, Ran BJ, et al. Adaptive multiple cues integration for robust outdoor vehicle visual tracking. In: The 34th Chinese Control Conference; July 28–30, Chengdu, China; 2015. pp. 4913-4918

[5] Wang L, Ouyang W, Wang X, et al. Visual tracking with fully convolutional networks. In: IEEE International Conference on Computer Vision. IEEE; 2015. pp. 3119-3127

[6] Nam H, Han B. Learning multi-domain convolutional neural networks for visual tracking. Computer Vision and Pattern Recognition. 2016:4293-4302

[7] Olshausen BA, Field DJ. Sparse coding with an overcomplete basis set: A strategy employed by V1? Vision Research. 1997;**37**(23):3311-3325

[8] Rehn M, Sommer FT. A network that uses few active neurones to code visual input predicts the diverse shapes of cortical receptive fields. Journal of Computational Neuroscience. 2007;**22**(2):135-146

[9] Zheng Y et al. Deep learning and its new progress in object and behavior recognition. Journal of Image and Graphics. 2014;**19**(2):175-184

[10] Makhzani A, Frey B. k-sparse autoencoders. In: International Conference on Learning Representations, ICLR, 2014. pp. 235-241

[11] Fan J et al. Human tracking using convolutional neural networks. IEEE Transactions on Neural Networks. 2010;**21**(10):1610-1623

[12] Jin J, Dundar A, Bates J, et al. Tracking with deep neural networks. In: Conference on Information Sciences and Systems. 2013. pp. 1-5

[13] Hong S, You T, et al. Online tracking by learning discriminative saliency map with convolutional neural network. Computer Science. 2015. pp. 597-606

[14] Wang L, Liu T, et al. Video tracking using learned hierarchical feature. IEEE Transactions on Image Processing. 2015;**24**(4):1424-1435

[15] Wang N, Li S, Gupta A, et al. Transferring rich feature hierarchies for robust visual tracking. arXiv preprint arXiv:1501.04587, 2015

[16] Doulamis N, Voulodimos A. FAST-MDL: Fast adaptive supervised training of multi-layered deep learning models for consistent object tracking and classification[C]. In: IEEE International Conference on Imaging Systems and Techniques (IST), Chania; 2016. pp. 318-323

[17] Guan H, Xue XY, An ZY. Advances on application of deep learning for video object tracking. Acta Automatica Sinica. 2016;**42**(6):834-847

[18] Zhang P, Zhuo T, Huang W, et al. Online object tracking based on CNN with spatial-temporal saliency guided sampling. Neurocomputing. 2017:**17**(4):1-13

[19] Wang N, Yeung DY. Learning a deep compact image representation for visual tracking. In: International Conference on Neural Information Processing Systems; 2013. pp. 809-817

[20] Zhou X, Xie L, Zhang P, Zhang Y. An ensemble of deep neural networks for object tracking. In: IEEE International Conference on Image Processing; 2014. pp. 843-847

[21] Cheng S, Sun JX, et al. Target tracking based on incremental deep learning. Optics and Precision Engineering. 2015;**23**(4):1161-1170

[22] Cheng S, Cao YG, Sun JX, et al. Target tracking based on enhanced flock of tracker and deep learning. Journal of Electronics and Information Technology. 2015;**37**(7):1146-1153

[23] Cheng S, Sun JX, et al. Target tracking based on multiple instance deep learning. Journal of Electronics and Information Technology. 2015;**37**:12

[24] Bo D, Hou ZQ, Yu WS, et al. Local patch tracking algorithm based on stacked denoising autoencoder. Control Theory & Applications. 2017;**34**(6):829-836

[25] Hua W, Mu D, Guo D, et al. Visual tracking based on stacked denoising autoencoder network with genetic algorithm optimization. Multimedia Tools & Applications. 2017;**2**:1-17

[26] Torralba A et al. 80 million tiny images: A large data set for nonparametric object and scene recognition. IEEE Transactions on Pattern Analysis and Machine Intelligence. 2008; **30**(11):1958-1970

[27] Richard MD et al. Neural network classifiers estimate Bayesian a posterior probabilities. Neural Computation. 1991;**3**(4):461-483

[28] Wu Y, Lim J. Online object tracking: A benchmark. In: IEEE Conference on Computer Vision and Pattern Recognition. 2013;**9**(4):2411-2418

[29] Ahuja N. Robust visual tracking via multi-task sparse learning. In: IEEE Conference on Computer Vision and Pattern Recognition. 2012;**157**:2042-2049

[30] Henriques JF, Rui C, et al. Exploiting the circulant structure of tracking-by-detection with kernels. Computer Vision. 2012;**7575**:702-715

[31] Masci J, Meier U, Dan C, et al. Stacked convolutional autoencoders for hierarchical feature extraction. Artificial Neural Networks and Machine Learning-ICANN 2011. In: International Conference on Artificial Neural Networks; 2011. pp. 52-59

[32] Xin J, Du X, Zhang J. Deep learning for robust outdoor vehicle visual tracking. In: IEEE International Conference on Multimedia and Expo (ICME 2017); July 10–14, 2017. Hong Kong, China. pp. 613-618

4

Traditional and New Types of Passive Flow Control Techniques to Pave the Way for High Maneuverability and Low Structural Weight for UAVs and MAVs

Mustafa Serdar Genç, Kemal Koca,
Hacımurat Demir and Halil Hakan Açıkel

Abstract

Prevailing utilization of airfoils in the design of micro air vehicles and wind turbines causes to gain attention in terms of determination of flow characterization on these flight vehicles operating at low Reynolds numbers. Thus, these vehicles require flow control techniques to reduce flow phenomena such as boundary layer separation or laminar separation bubble (LSB) affecting aerodynamic performance negatively. This chapter presents a detailed review of traditional passive control techniques for flight vehicle applications operating at low Reynolds numbers. In addition to the traditional methods, a new concept of the pre-stall controller by means of *roughness material*, *flexibility* and *partial flexibility* is highlighted with experimental and numerical results. Results indicate that passive flow control methods can dramatically increase the aerodynamic performance of the aforementioned vehicles by controlling the LSB occurring in the pre-stall region. The control of the LSB with new concept pre-stall control techniques provides lift increment and drag reduction by utilizing significantly less matter consumption and low energy. In particular, new types of these methods presented for the first time by the chapter's authors have enormously influenced the progress of separation and LSB, resulting in postponing of the stall and enhancing the aerodynamic performance of wind turbine applications.

Keywords: energy harvesting recovery, passive flow control techniques, partial flexibility, pre-stall control mechanisms

1. Introduction

Boundary layer transition and separation phenomena have been researchable topics for over 100 years, but there are still many open essential issues and practical challenges containing their controls. It is predicted that the fuel cost of a commercial aircraft could be saved to 8% if the transition phenomenon over its wing could be delayed to 50% [1]. These flow phenomena commonly occur at low Reynolds numbers (Re) at which laminar flow is dominant. A laminar boundary layer can separate from the solid surface when the adverse pressure gradients (APGs) play a preponderant role. Transition phenomenon in the separated region is caused by the separated shear layer, and then the turbulent reattachment starts to occur because of energized vorticial structures. The region between separation and reattachment points is called as laminar separation bubble [2, 3] (LSB), which negatively affects the aerodynamic performance. LSBs can form in many aeronautical applications operating in low Re regime of less than 1×10^6 and angle of attack (AoA) of less than stall angle, such as high latitude aircrafts, micro air vehicles (MAVs), multielement airfoil configurations, unmanned aerial vehicles (UAVs), wind turbine, and low-pressure turbine blades.

As stated above, the aerodynamic efficiency can be severely decreased by LSBs by reducing the lift and increasing the drag forces. Besides this, it causes the increment of unsteadiness and noise, especially for wind turbine applications. Regarding a better understanding of LSBs' topology, they can be categorized as 'short' and 'long' bubbles. Aerodynamic researchers recommended a few parameters so that LSBs could be classified whether they are 'short' or 'long' [3–5]. Assessment of effects of either 'short' or 'long' bubbles on the pressure distribution can be the best option and intuitive way. If the pressure field is drastically affected by strong downstream and upstream impacts, it is 'long'. But, if the LSB causes the local and limited impacts on the pressure distribution, it is 'short'. LSBs have an unsteady separated shear layer because of the small disturbances and these disturbances cause the vorticial structures to grow rapidly. The separation and reattachment points are affected by these structures and they inherently cause the shape and size of the separated region to change. Therefore, the laminar to turbulent transition and the instability procedures significantly affect the unstable characteristics of LSBs and the mean flow topology [6–8]. Toward this end, a comprehensive understanding of the concerning physical mechanisms of separation and transition procedures is necessary. This may be concluded with beneficial separation prediction tools, which can result in the development of geometrical structures hydrodynamically and aerodynamically. The probability of active and passive flow control techniques, which cause the negative effects of separation and LSBs to suppress (or at least diminish), may also be revealed by means of these understandings.

The objective of this study is to elucidate the traditional passive control techniques for wind turbine applications operating at low Reynolds number regimes. Besides the explanation of traditional low Reynolds number flow control methods such as VGs etc., new concept pre-stall control mechanisms such as roughness material, flexibility and partially flexibility as mentioned in detailed experimental studies will be enormously highlighted.

2. Low Reynolds number aerodynamics

The aerodynamic performances of airfoils in low Reynolds number flows considerably relate to a major range of engineering applications. The low Reynolds number flows are described that the viscous forces within fluid gain dominant characterization compared with inertial forces. Hereby, boundary layer physics such as laminar boundary layer separation, reattachment, and transition phenomena can frequently occur at low Reynolds number flows, affecting the performance of airfoils with the important changes of lift and drag forces.

2.1. Transition modes

2.1.1. Natural transition

Given there was an inflection point in the velocity profile, it was proposed that all boundary layer profiles were unstable according to inviscid stability theory. Prandtl [9] later explained a physical prediction of transition and then Tollmien [10] proved it mathematically that viscous instability waves (often identified as Tollmien-Schlichting (hereinafter TS) waves) could cause a laminar boundary layer to destabilize. But these findings were not accepted by aerodynamic researchers until the study performed by Schubauer and Skramstad [11], since the early experimental investigation for transition prediction had large free-stream turbulence level.

Now, the free-stream turbulence level is accepted as low when it is less than 1% (<1%) [12]. In that flow case, a laminar boundary layer becomes linearly unstable when the critical Reynolds number is increased, showing TS waves that have started to grow.

The sketch of the natural transition process is indicated in **Figure 1**. The flow in the area indicated by number 1 is laminar. In the area denoted by number 2, TS waves start to grow. After that point, the transition to turbulence may not be concluded every time due to the slow growth of TS waves. In aerodynamic literature, the regular transition starts to occur after the nonlinear waves have taken place. It can be said that the transition to turbulence is inevitable

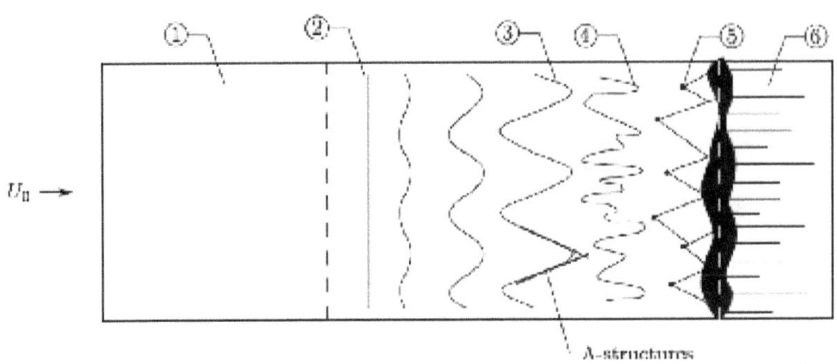

Figure 1. The sketch after Kurelek [13], Bertolotti [14] and Schlichting and Gersten [15] with regard to top view of simplified flat plate boundary layer transition. (1) Laminar flow, (2) TS waves, (3) three-dimensional waves and Λ-structure formation, (4) vortex breakdown, (5) turbulent spots formation, (6) turbulent flow.

after Λ-structures and three-dimensional disturbance defined as nonlinear waves in the area of number 3 and 4. Once those structures form, spots spread in all directions, resulting in the existing turbulent boundary layer as shown in numbers of 5 and 6, respectively.

2.1.2. Bypass transition

Another type of transition is bypass transition. In this transition phenomenon, 2D instability cases of natural transition shown in **Figure 2** are bypassed when the boundary layer on a flat plate under free-stream turbulence intensity is larger than 1%. But, the explanation of bypass transition has still a mystery when free-stream turbulence level is 1%. Therefore, aerodynamic researchers accept as the boundary among natural and bypass transition as free-stream turbulence level is 1%. As mentioned before, spanwise vorticity and 3D breakdown are bypasses in bypass transition and the turbulence spots are directly produced in the boundary layer.

2.1.3. Separated flow transition

Transition to turbulence can exist in the free shear layer when the laminar boundary layer is separated from a solid surface due to sharp, blunt or rounded leading edges. As explaining physically, the boundary layer is able to overcome the effect of adverse pressure gradients at high Reynolds numbers. But, especially at low Reynolds numbers, the laminar boundary layer cannot overcome the adverse pressure gradient due to lack of momentum in the boundary layer and it separates from the surface. This flow separation can cause the transition in the free shear layer. Moreover, after a while, the separated flow may reattach to the surface, resulting in the formation of the laminar separation bubble.

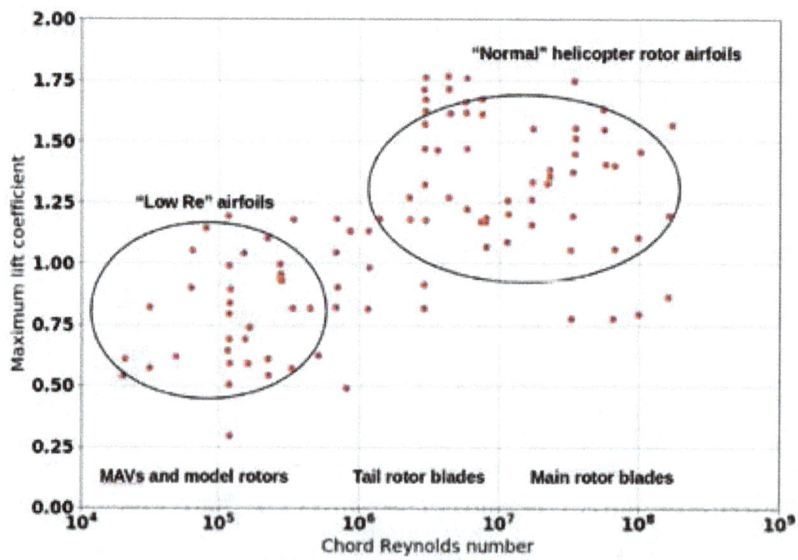

Figure 2. The different rotary wing applications according to variation in chord Reynolds number [18].

2.1.4. *Wake-induced transition*

This type of transition phenomenon can generally be observed in turbomachinery flows because flows on the blade rows are subjected to periodically impinging wakes coming from the preceding blade rows [16]. In aerodynamic literature, it is suspicious whether enhanced turbulence in the wake or the free-stream level triggers the transition phenomenon. Thus, this type of transition is occasionally distinguished from bypass transition phenomenon and is referred to as wake-induced transition.

2.1.5. *Reverse transition*

Also known as re-laminarization, the reverse transition is the transition of turbulent to laminar flow. This mode of transition can highly be observed because of high flow acceleration in the regions where favorable pressure gradients play a dominant role. It can be noted that the regions that have favorable pressure gradient generally occur on an airfoil near the leading edge of suction surface and close to the trailing edge of pressure surface. According to the notification carried out by Narasimha and Sreenivasan [17], turbulence dissipation, surface mass transfer, and thermal effects can cause possible reverse transition.

Regarding the different rotary wing applications, the variation of maximum lift coefficient ($C_{L, max}$) with the chord-based Reynolds number range is shown in **Figure 2** [18]. The fluid flows over airfoils at especially chord-based Reynolds number of 10^4 to 10^5 are more sophisticated due to the dominant character of viscous effects. Despite most regarding studies performed by aerodynamic researchers, low Reynolds number aerodynamics still have researchable potential. This is because of the following: (i) the separated laminar boundary layer because of adverse pressure gradients (APGs) is sensitive; (ii) transition region is too broad, resulting in more unsteady behavior; (iii) short and long separation bubble formations emerge with these APGs and inadequate momentum in flow; and (iv) susceptible role can be played because of surface conditions free-stream turbulence.

Due to these types of flow phenomena mentioned above such as laminar boundary layer separation or LSBs, the detection of flow separation is too important. Flow separation can cause the aerodynamic stall leading to undesired dynamic or static loading statements with a decrease in the lift and an increase in drag for airfoils operating at low Reynolds number ranges. Therefore, flow control methods have been developed to mitigate (to even suppress) their detrimental influences in terms of aerodynamic performances. The flow control mechanism can be divided into two categories as passive and active flow control. The essential difference between active and passive flow control techniques is that some sort of energy input is necessary for active flow control to manipulate the flow, while passive flow control methods manipulate the flow by not requiring any exterior energy resources. These two control methods have advantages and disadvantages compared to each other. One drawback of passive flow control technique can be that it cannot be switched on or off whenever users need. But most aerodynamic researchers have recently preferred the passive flow control methods to provide technologically and economically efficient solutions as long as it does not ensure any undesirable situations except for its design conditions, because they are the quickest solution to implement processes and less expensive. In spite of the advantages and disadvantages of these two control techniques, they have been tested and researched by aerodynamic researchers with the aim of alleviating the stall effects and enhancing the performance of the overall airfoil.

3. Passive flow control techniques

These types of flow control methods generally improve the condition of flight vehicles by manipulating the flow characteristics in the boundary layer, because the flow manipulation is an efficient way to control mixing in the separated shear layer. So far, aerodynamic researchers have utilized these techniques experimentally or numerically in their studies. A detailed explanation of these passive flow control techniques has been presented as follows:

3.1. Vortex generators

The vortex generators (VGs) as depicted in **Figure 3** [19] are the most effective and simplest passive flow control devices that are widely preferred and utilized on wind turbine blades by aerodynamic researchers in order to prohibit and suppress flow separation caused by APGs. VG examples are not limited to airfoil [20], and they can also utilize the devices such as bluff bodies [21], noise reduction [22], wind turbines [23], swept wings [24], and heat exchangers [25, 26], just to name a few. VGs, which were first investigated by Taylor [27], are generally small plates having rectangular or triangular shapes. They can be mounted on the surface where desired to flow control at an angle of the incoming flow. They are used to decrease (to even suppress) the boundary layer separation, which is caused by APGs and turbulence effects [28]. The slower moving boundary layer is energized by VGs in conjunction with high momentum fluid in the outer part of boundary layer and in the free stream [29], resulting in reducing the drag force [30] and increasing the lift force [31, 32].

Regarding the optimization of VGs, many aerodynamic researchers have investigated the VG's parameters to obtain optimal impacts on fluid flow. As shown in **Figure 4** [33], it can be said that the important parameters are height (h), cropped edge length (b), vane length (L),

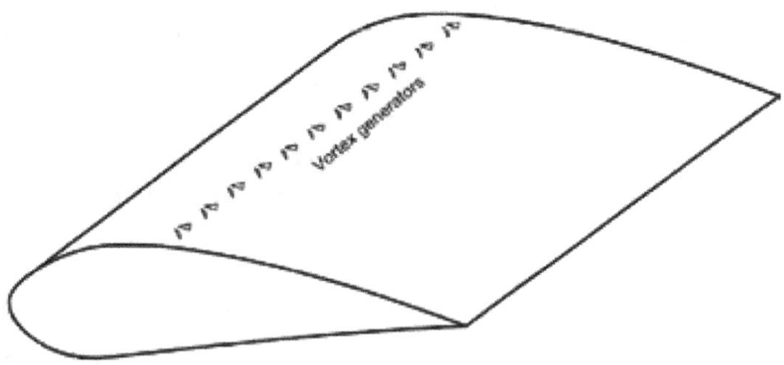

Figure 3. A sketch of vortex generator rows [19].

type, shape, pattern, long (D) and short (d) gaps among the vanes, size, location, and inflow angle (β). In addition to these important parameters, VG configurations termed as counter rotational and corotational also play critical roles in terms of rotational directions of vortices formed by VG pairs [34].

3.2. Leading-edge slats

The leading-edge slats, which were known as a passive flow controller by delaying the flow separation, were initially presented by Handley Page [35] in Great Britain and it was first utilized for an aircraft [36]. As illustrated in **Figure 5** [19, 37], the flow in space between the main body and slat is augmented and accelerated with either large vortices or multiple smaller vortices. Large vortices moving from slat's midspan to its edge can occur at lower Reynolds

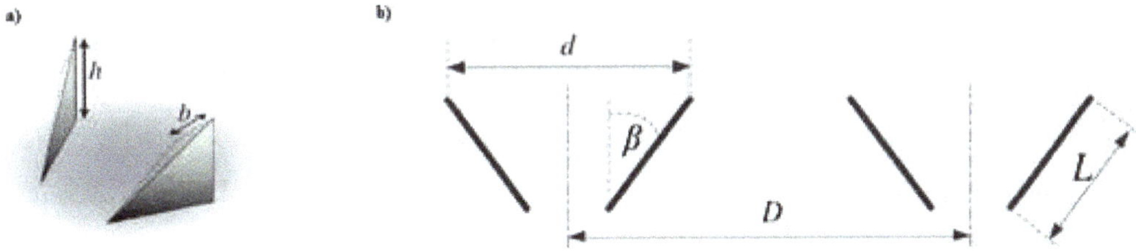

Figure 4. (a) Isometric perspective and (b) planar sketch of the VG row [33].

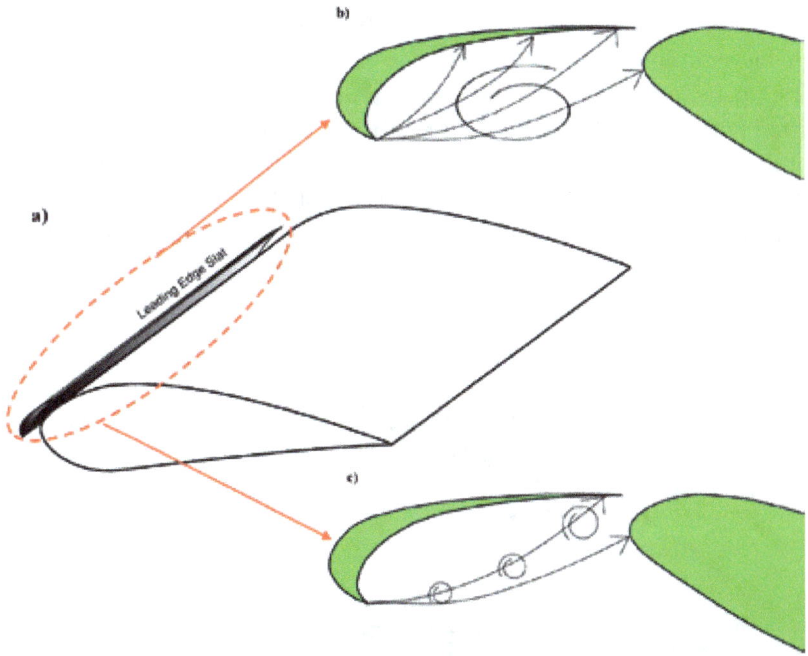

Figure 5. (a) Configuration of leading-edge slat [19] and (b and c) its planar view [37].

number, whereas smaller vortices can be observed at higher Reynolds number. Accelerating flow with leading-edge slats gains kinetic energy and momentum to the boundary layer, resulting in delaying of stall phenomenon [38]. In the literature, there are three types of slats: (i) fixed slat [39], (ii) retractable slat [40], and (iii) Kruger flap [41]. Recently, Genç et al. [42] have investigated NACA2415 airfoil with NACA22 leading-edge slat experimentally and computationally. Their computational results indicated that experimentally stated LSB was correctly estimated. Moreover, delaying of the stall phenomenon was obtained by means of experimental investigation, resulting in providing the maximum lift coefficient of 1.3.

3.3. Flow vanes

The flow vane, which is concepted by Pechlivanoglou [19] and can be utilized as a power regulator and stall controller at wind turbines, is an undiscovered item of the wind turbine blades. As seen in **Figure 6**, the flow vanes have relatively smaller chord length than the main body and this additional aerodynamic profile can be positioned over the suction surface of airfoils. The space among the flow vane and main body is closely equal to the chord length of the flow vane.

3.4. Leading-edge serrations

A passive flow control method entitled as leading-edge serration as shown in **Figure 7** [43, 44] is inspired by the morphology of humpback whales [45]. This bioinspired technique has recently been investigated for different purposes experimentally or numerically. Wang and Zhuang [46] designed a modified wind turbine blades with sinusoidal wave serrations employed on the leading edge to control the boundary layer separation. Their numerical results indicated that the leading-edge serration suppressed the flow separation with the generation of the counter-rotating vortex pairs, especially at high AoA. Cai et al. [47] also numerically investigated a modified airfoil with a single leading-edge protuberance at low Reynolds number. The results showed that the stall angle reduced at the modified airfoil. Furthermore, the pre-stall performance of the modified airfoil decreased, whereas post-stall characteristics were increased. Moreover, an experimental study performed by Wei et al. [48] expressed the hydrodynamic

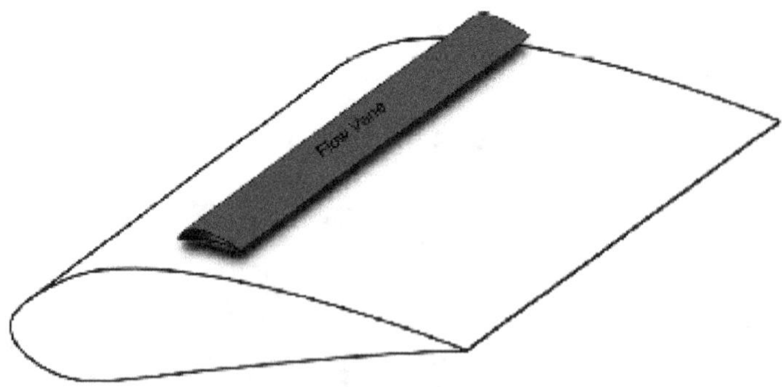

Figure 6. Representation sketch of the flow vane [19].

characteristics of hydrofoils with leading-edge tubercles at Reynolds number of 1.4×10^4. Their visualization results based on particle tracking revealed that the effects of flow separation were declined with the use of leading-edge tubercles.

3.5. Slotted airfoils

Slots (generally known as a narrow rectangular channel along spanwise of a wind turbine blade) are one of passive flow control methods and flow control is ensured by changing the flow velocity over the airfoil. The principle of a slotted airfoil is that flow velocity increases at the slot exit after interior flow passes within the airfoil. This increment of flow velocity at the slot exit causes the streamlines to disrupt, resulting in creating the flow separation. The flow separation occurred over the airfoil means velocity reduction. This reduction in flow velocity enables the local pressure underneath the airfoil to increase, resulting in producing more lift. **Figure 8** illustrates the slot geometric characteristics performed by Belamadi et al. [49].

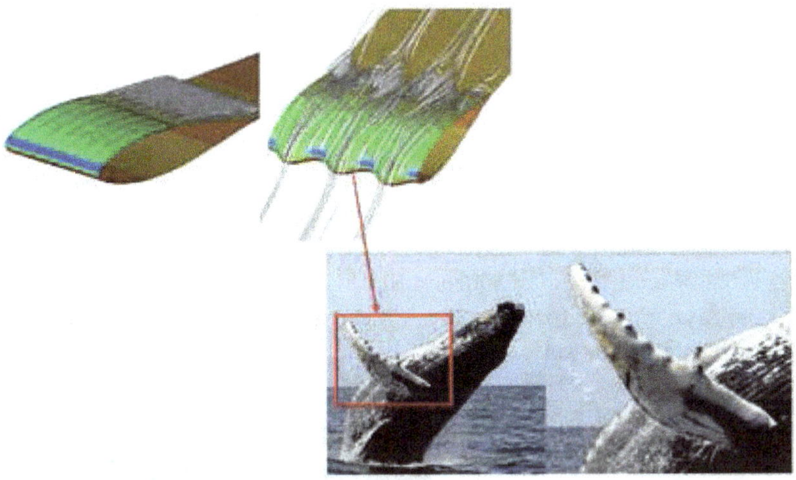

Figure 7. Schematic view of a wind turbine blade with modified models via serrations [43, 44].

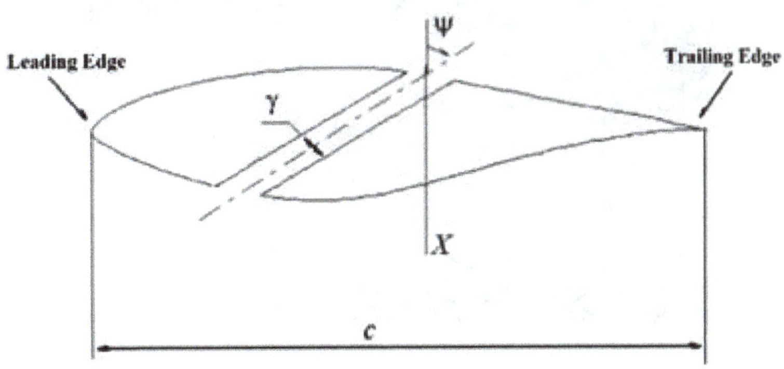

Figure 8. Schematic demonstration of a slotted airfoil [49].

Symbols of c, X, γ, and ψ mean chord of the airfoil, the slot position, the slot width, and angle between slot axis and chord normal, respectively. Numerical results conducted by Belamadi et al. [49] indicated that stall phenomenon on S809 airfoil at an angle of attack of 20° was completely eliminated by creating a nozzle effect over the airfoil, ensuring the extra kinetic energy (inherently extra momentum) to suction surface. Based on experimental and numerical results obtained by Beyhaghi and Amano [50], an increment of the lift coefficient by 30% was ensured without conceding any drag force.

3.6. Leading-edge microcylinder

The leading-edge microcylinders, which are used as passive control technique for boundary layer flow separation, are shown as a whole and enlarged view of mesh domain in **Figure 9** [51]. Regarding the principle of leading-edge microcylinder, velocity over suction surface of the airfoil can be accelerated by them and thus the Kelvin-Helmholtz instability of fluid flow is decreased. As concerning literature studies performed in advance, Luo et al. [52] designed a microcylinder and used in front of the leading edge of NACA0012 airfoil in order to ensure stall delay and decrease the flow separation zone. Based on the numerical calculation carried out by Wang et al. [53], an increment of 27.3% at blade torque was obtained by positioning a microcylinder with a suitable diameter in front of the leading edge.

3.7. Gurney flaps

Gurney flap is a small boundary layer passive control method and it can be easily mounted at the trailing edge of an airfoil. The Gurney flap with 2% of the chord length of the airfoil can affect the aerodynamic performance by increasing the lift coefficient by 0.4. Moreover, the lift-to-drag ratio of the airfoil can be nearly improved by 35% [54]. Flow progress and

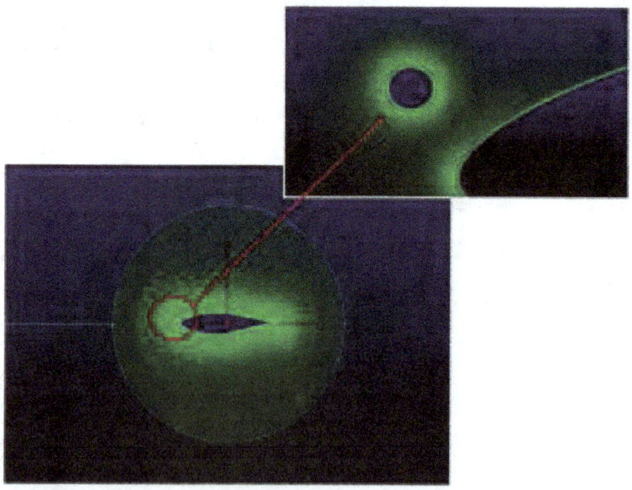

Figure 9. Whole and enlarged view of mesh domain for a leading-edge microcylinder [51].

mechanism of lift increment by the Gurney flap were explained by Liebeck [55]. As depicted in **Figure 10** [55], a pair of counter-rotating vortices composed at downstream of the Gurney flap creates a low-pressure zone within. This low-pressure region makes the flow to increase over the suction surface, resulting in rising to an increase in the suction pressure. On the other hand, the flow velocity is reduced at upstream of Gurney flap by anticlockwise vortices and pressure at the pressure surface is increased. Consequently, the variation of pressure distribution between surfaces leads to an increment of lift force. The unsteady flow characteristics, especially at low Reynolds numbers, may be mitigated and suppressed with the use of Gurney flap. Based on the transient two-dimensional numerical simulations performed by Zhu et al. [56], the adaptive Gurney flap was compared with the fixed Gurney flap and the greater energy harvesting efficiency was obtained when the adaptive Gurney flap was selected for the oscillating wing. Shukla and Kaviti [57] numerically investigated four symmetric NACA airfoils in conjunction with a dimple, Gurney flap and combination of both dimple and Gurney flap at Reynolds number of 3.6×10^5. Their results indicated that better aerodynamic performance was obtained at NACA0021 airfoil with a combination of both dimple and Gurney flap when the angle of attack was 12°.

3.8. Self-activated deployable flap

Recently, aerodynamic researchers have focused on a technique to palliate the adverse effect of flow and increase the lift coefficient by inspiring the biological flows from birds' wings. Therefore, a self-activated spanwise flap near the trailing edge of the airfoil as a biomimetic device for control of flow separation has been developed as seen in **Figure 11** [58]. Regarding the principle of the self-activated flap, it starts to pop-up because of backflow to cope with critical conditions of flight when flow separation occurs on the suction surface of the wing. Thus, the wing can be prevented from an abrupt increase in the angle of attack because of perching maneuvers or gusts. Rosti et al. [59] investigated physical mechanism of the flow field over NACA0020 airfoil having an elastically mounted flap at Reynolds number of 2×10^4. It was founded that these flaps could overcome the effect of dynamic stall breakdown causing the abrupt lift loss. In addition, a more positive aerodynamic response such as increasing of lift amount was obtained during the ramp up motion of movable flap. Arivoli and Singh [60] and Schluter [61] also studied self-activated deployable flaps. Their results demonstrated that deployable flaps played crucial roles in terms of aerodynamic performance even though they had a little effect in Reynolds number and heavy stall conditions.

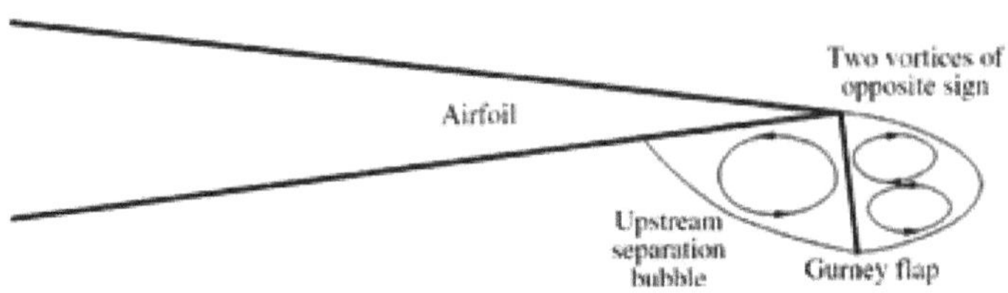

Figure 10. Flow progress at trailing-edge airfoil having a Gurney flap [55].

3.9. Airfoils with cavity

One of the passive control techniques is the application of a cavity mounted on a thick airfoils' suction surface that is taken from original Kasper's wings [62]. The principle of this concept is to create a convenient pressure gradient when two counter-rotating vortices inside the cavity are trapped. Furthermore, these trapped vortices over the suction surface not only ensure an extra low-pressure region but also cause a lower drag to produce. Thus, this method has recently gained interests among aerodynamic researchers. Olsman and Colonius [63] investigated an airfoil with a cavity at Reynolds number of 2×10^4 and different angles of attack ranging from 0° to 15° as seen in **Figure 12**. Their results revealed that stall phenomenon was delayed by means of counter-rotating separated flows, resulted in reduced flow separation region. A detailed numerical study regarding the aeroacoustics of NACA 0018 cavitied airfoil was reported by Lam and Leung [64] at Reynolds number of 2×10^4 and Mach number (Ma) of 0.2.

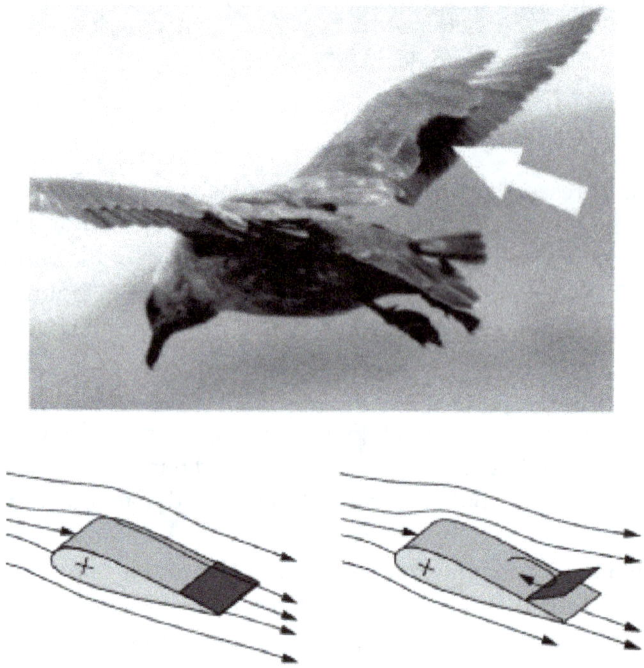

Figure 11. Illustration of a movable flap at a high angle of attack: seagull at landing (top picture) and sketch in principle (bottom picture) [58].

Figure 12. The vorticity contour plot over the airfoil with a cavity [63].

The presence of cavity caused the lift-to-drag ratio to increase. Moreover, cavitied airfoil produced less acoustic power, making it a noiseless airfoil design at low Reynolds number regimes.

3.10. Roughness material

The control technique with the roughness material, which is one of the fundamental objectives of the chapter, can passively control the flow over wind turbine blade operating at low Reynolds number ranges. Regarding the development of vortex structures at the wake of VG applications [65], the flow is re-energized with the vortices produced by miniramps as denoted in **Figure 13**. Furthermore, the flow is inherently gaining momentum by means of that passive flow controller. Therefore, the separated flow because of adverse pressure gradients that occurred generally at leading edge of airfoils may be suppressed with re-energized flows, resulting in the occurrence of more stable flow characteristics without boundary layer separation.

As occurred in VG applications, the flow control method by means of roughness material is performed with similar ways by intervening the flow. Vortex sheds produced by roughness cause the flow in the boundary layer to gain more energy as seen in **Figure 14** [66]. Energized flow hinders the boundary layer flow separation and it ensures the flow to move along the airfoil surface by attaching. The vortex sheds can be used for a few different purposes over the surface of airfoils. For instance, the vortex sheds, which were used for recognition of flow phenomena at the study presented by Koca et al. [67, 68], gave the momentum to flow, resulting in lift recovery and even less vibration or noise for wind turbine blades.

Regarding identifying the role of roughness material on the flow characteristics over roughened NACA 4412 airfoil as indicated in **Figure 15**, investigations based on the force measurement, the smoke-wire, hot-film sensor (glue-on type), and hot-wire experiments have been performed by Genç et al. [8, 69]. The purpose of the experimental study was to determine the LSB and transition phenomena over uncontrolled NACA 4412 airfoil in detail and then was to observe how sandpaper as a roughness material affected the flow topology.

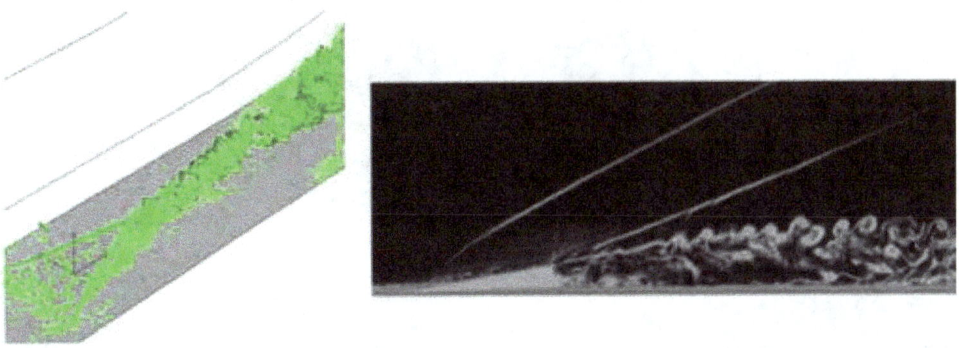

Figure 13. Vortex structures at the wake of VGs [65].

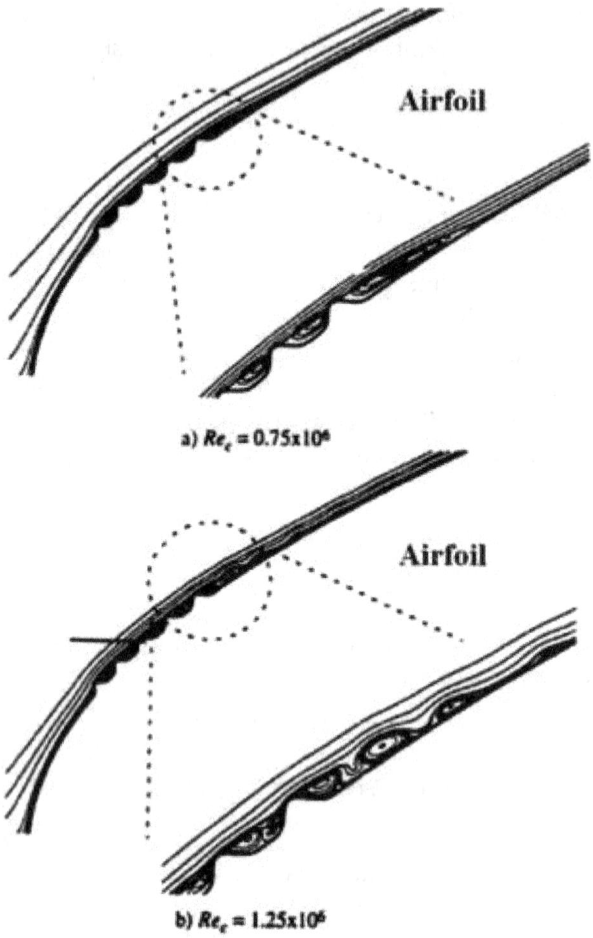

Figure 14. Comparison of ¼″ roughness height and vortex shedding characteristics at different Re_c numbers [66].

Figure 15. Representation sketch of the roughened airfoil.

The results, which were obtained from smoke-wire experiment and hot-film sensor, showing an integrated graph were denoted in **Figure 16** [8]. The streamlines obtained from smoke-wire experiment clearly revealed that LSB occurred between x/c = 0.3 and x/c = 0.7 for uncontrolled

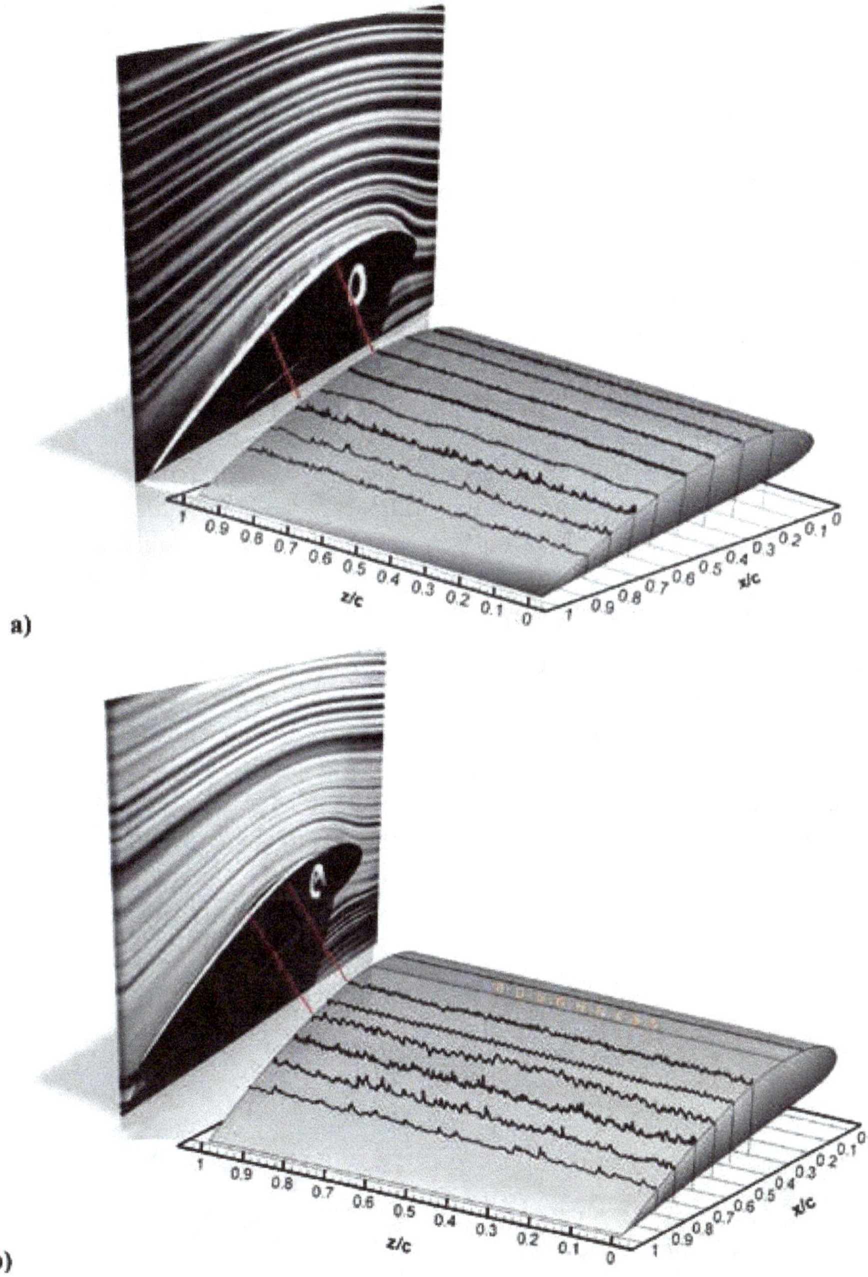

Figure 16. Comparison results of two different experiments at Reynolds number of 5×10^4 and $\alpha = 8°$: (a) k/c = 0 (uncontrolled airfoil) and (b) k/c = 0.003 [8].

airfoil, while it was seen between x/c = 0.3 and x/c = 0.5 for the roughened airfoil. It means that using sandpaper causes LSB's size to shrink enormously. As physically speaking, the undulations acquired from voltage values, which were predefined how to obtain in Ref. [8], started to increase after x/c = 0.3, meaning the transition inception and separation point due to adverse pressure gradients in **Figure 16(a)**. However, the amount of undulations at x/c = 0.5 was less than that at x/c = 0.3, because small eddies having less energy in the aft portion of LSB caused the undulations amount to reduce. After x/c = 0.5 point, the obvious increment in undulations indicated that the flow in the boundary layer was fully turbulent because of energized vortices.

Figure 17 [8] shows a combination graph consisting of numerical and experimental results for a roughened airfoil with k/c = 0.006 at Reynolds number of 5×10^4 and $\alpha = 8°$. At first glance, APG exhibits a dominant role on flow and it causes the flow to separate from the airfoil surface of x/c = 0.3 as depicted in the flow visualization graph. Then, the flow reattaches to solid surface nearly at x/c = 0.6 by gaining momentum by means of roughness material. Same flow phenomena like boundary layer separation, reattachment and LSB are shown and proved with streamlines and C_p curves obtained from the numerical study. The peak point among separation (referred to as S) and reattachment (referred to as R) points reveal the LSB in C_p curve. The trend of Cp curve is almost constant after separation point due to the presence of dead air region having as negligible as less flow phenomenon. The position of LSB is between x/c = 0.3 and x/c = 0.6 as shown in the smoke-wire experiment result. Besides, a mild peak at x/c = 0.5 indicates the transition point over the airfoil surface.

In addition to the results mentioned above, two more important results were obtained from aerodynamic force measurement results as seen in **Figure 18**. First, the stall phenomenon because of flow separation was pronouncedly postponed in **Figure 18(a)**. Second, lift coefficient (C_L) in **Figure 18(b)** increased with the use of roughness material, resulting in enhancement of aerodynamic performance of airfoil. Moreover, it was clearly seen that roughness material gave good results, especially in the pre-stall region. Thus, the

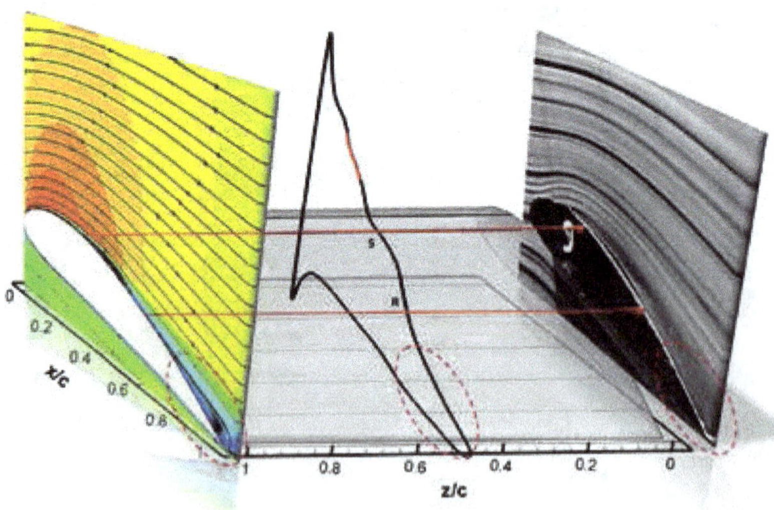

Figure 17. The combined results obtained from the numerical and smoke-wire result for the roughened airfoil with k/c = 0.006 at Reynolds number of 5×10^4 and $\alpha = 8°$ [8].

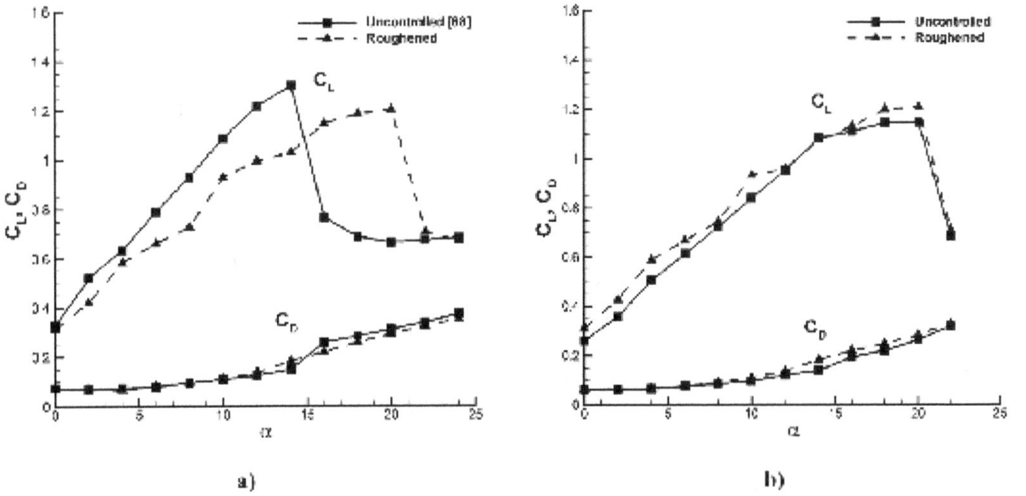

Figure 18. Force measurement results at k/c = 0.006: (a) Re = 7.5×10^4 and (b) Re = 1×10^5 [8].

roughness material was firstly entitled as "the pre-stall flow control mechanism" in aerodynamic literature by authors in Ref. [8].

3.11. Flexible membrane wings

Another essential objective of the chapter in terms of the passive flow control device is doing a detailed survey on flexible membrane wings. The requirement for improving the flight capabilities of MAV and UAV leads to increasing concern in biologically inspired wings. It is well known that the wings of flying animals such as bats resemble a thin membrane-like material with a fixed leading edge and free, scalloped trailing edge that can be easily complied with the flow environment. Moreover, they can regulate the wing planform for a specific flight condition and their flight can be qualified by immensely unsteady and three-dimensional wing motions. A membrane wing is better able to tailor the atmospheric disturbances and makes the vehicle easier to fly [70, 71]. In other respects, the efficacy of the membrane comes from the ability of passive control through the flight as well as decreasing the weight of the wing [70]. Smith [72] paraphrased the emphasis on flexibility and wing stiffness in modeling the flapping motion and generation of the resultant force. A summary presentation study of the aerodynamics of micro air vehicles operating at low Reynolds numbers was carried out by Mueller and DeLaurier [73]. In order to come up with the negative effects of LSB for improving aerodynamic performance, researchers utilized flexible membrane wings for numerous practices such as hang glider, microlight, and UAVs and MAVs. An experimental investigation about time-dependent LSB formation on AR = 3 wing was conducted, and it was seen that in the membrane wings at low Re numbers the LSB was more prevalent. Leading-edge separations were influenced via both Reynolds number and leading-edge vortices occurring because of the separation bubbles led to time-dependent alterations on the vibration of the wing [74, 75]. At low AR, tip vortices delayed stall, exclusively at low Re numbers owing to affecting flow on the wing and separation bubble [76] and analysis of the instant deformation found out spanwise and chordwise, which were due to the shedding of leading-edge vortices' farther tip vortices [77].

Rojratsirikul et al. [78, 79] searched flow and deformation characteristics of membrane wings with low aspect ratio via velocity and deformation measurement. They found membrane oscillations in second chordwise mode at higher incidences. The dynamic of membrane wing can be altered with excess length [80, 81] and support [82] of the wing. Genç [80] studied on a membrane wing with excess length. The results depicted that camber of membrane wing induced the separated flow; therefore, small separated regions were seen. Besides, Greenhalgh et al. [81] observed that increasing excess length caused to reduce separation incidence, and hysteresis interval concluded a restricted working area for the highest excess lengths. Arbós-Torrent et al. [82] considered the effects of the geometry of front and aft of the wing on the aeromechanics of membrane wings. It was stated that average camber-like membrane fluctuations altered with respect to the geometry and size of both front and aft supports. Besides, the front and aft support having rectangular cross-section everlastingly provided further lift and deformations of mean camber compared with circular cross-sectional support. Galvao et al. [83] studied experimentally on the compliant membrane wings modeled based on mammalian flight mechanics. They showed that three-dimensional (3D) flow and tip vortices were ascendant. Furthermore, the deformation of compliant wings increased with both incidence and deformation increasing. Bleischwitz et al. [84] surveyed membrane wings aeromechanics in ground effect. Digital image correlation (DIC) and proper orthogonal decomposition (POD) were used for obtaining membrane vibrations. It was seen that fluctuation modes were adequate to hold 90% of all deformation energy closes to stall. Moreover, structural modes of spanwise were ensured in virtue of POD in lift increment areas. On the other hand, Hu et al. [85] executed a study on the flapping flexible membrane wings. It was seen that oscillation provided significant aerodynamic benefits in unsteady state regime. In other respects, it was concluded that generally the rigid wing had better lift capacity for flapping wings. The flexibility of the wing affected its aerodynamics positively [86] and membrane wings had an increase in maximum C_L during oscillating. Furthermore, an increase in reduced frequency led to an increase in maximum C_L. Membrane wings have a higher slope of lift and postponed stall [87]. Additionally, the membrane kinematics was closely relevant to membrane tension and free-stream velocity [88].

As previously mentioned, numerous studies have been performed for a better understanding of flexible wings and examined their effects in terms of aerodynamic performance. Herein, it is important to give sight for the conducted researches about flexible membrane wing, which are tabulated in **Table 1**. A common result could be said from all these studies that membrane wings had favorable characteristics such as a higher lift-to-drag ratio and a higher maximum lift coefficient when compared to an equivalent rigid wing from the aerodynamics point of view.

Unlike these studies, Demir [98] investigated the deformation that occurred on the flexible membrane wing surface and how it affected the LSB. Moreover, he examined how LSB affected the vibrations that occurred on the membrane surface, the distribution of the flow characteristics, as well as the fluid-structure interactions between the membrane and flow both

Author(s)	Type of wing	Type of study	Working range of Re number
Timpe et al. [71]	Rigid flat plate and membrane wings (AR = 4.3)	Experimental	5×10^4
Rojratsirikul et al. [79, 89, 90]	Rigid and 2D flexible membrane wing		5.31×10^4, 7.97×10^4, 10.6×10^4
Rojratsirikul et al. [78]	AR = 2 flexible membrane wing and delta wing		2.4×10^4–5.9×10^4
Hu et al. [85]	Cybird-P1® remote control ornithopter model		1×10^4, 2×10^4, 8×10^4
Tamai et al. [91]	Flexible membrane wings with different numbers of ribs (FM01, FM02, FM03, FM10)		7.5×10^4
Arbós-Torrent et al. [82]	Membrane airfoils with different geometries of LE and TE support		9×10^4
Bleischwitz et al. [84]	Rigid wing and AR = 2 membrane wing		5.6×10^4
Wrist et al. [92]	Membrane wing with NACA 2504, 4504, 6504, 4404, 4504, 4604, and 4506 frames		5.0×10^4
Attar et al. [93]	Membrane wing		1.37×10^4, 2.26×10^4, 3.63×10^4
Galvao et al. [83]	Membrane wing (AR = 0.92)		7×10^4–2×10^5
Viieru et al. [94]	Fruit fly (Drosophila)	Numerical	10^4–10^5
Hefeng et al. [95]	NACA0012 segmented flexible airfoil		1.35×10^5
Lian and Shyy [96]	Flexible wing (SD7003)		6×10^4
Gordnier and Attar [97]	AR = 2 flexible membrane wing	Both	2.43×10^4
Song et al. [88]	Rectangular membrane wing (AR = 0.9, 1.4, 1.8)	Both	7×10^4–2×10^5

Table 1. Summary of pioneering studies on flexible membrane wing.

experimentally and numerically. An experimental study was conducted by Demir and Genç [74] in order to examine time-dependent circumstance of flow on flexible membrane wing and they noticed that the size of LSB altered with time because of the indecisive flow features of the wing. The indecisive behavior upon the flexible membrane wing brought about various deformation modes to constitute at various angles of attack. The results of time-dependent flow visualizations for angles of attack of $\alpha = 12°$ and $\alpha = 10°$ for different time intervals are given in **Figure 19** [75] and **Figure 20** [75]. Time-dependent attitudes of LSB was obviously seen as analogizing obtained results at miscellaneous times between t = 0.08 s and t = 0.20 s. The bubble size enlarged at t = 0.16 s and then was smaller at t = 0.20 s for Re = 2.5×10^4 at $\alpha = 12°$, as seen in **Figure 19**. Additionally, as it is seen in **Figure 20**, the size of LSB enlarged until t = 0.12 s and then lessened at t = 0.16 s at $\alpha = 12°$ and Re = 5×10^4. For this purpose, it can be deduced that bubble size varied with time because of the indecisive flow characteristics of flexible membrane wing.

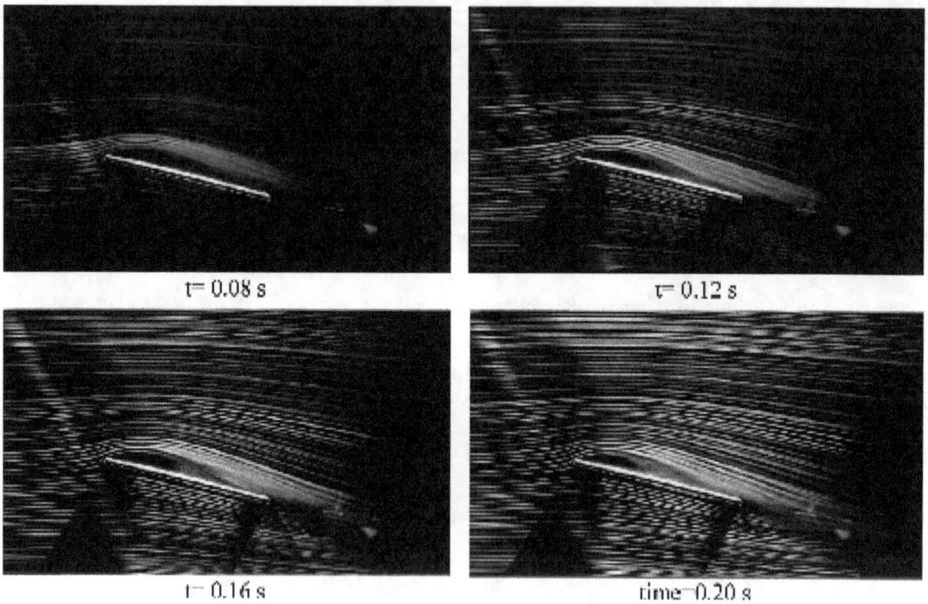

Figure 19. Smoke wire flow visualization result of AR = 3 flexible membrane wing at y/s = 0.4 for α = 12° and Re = 2.5 × 10^4 [75].

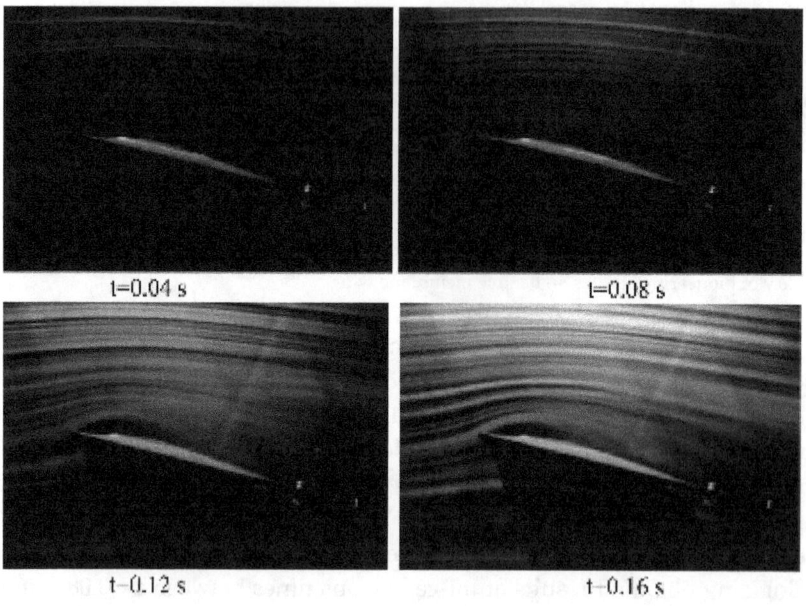

Figure 20. Smoke wire flow visualization result of AR = 3 flexible membrane wing at y/s = 0.4 for α = 10° and Re = 5 × 10^4 [75].

As seen in **Figure 21**, vibrational modes in the middle section of the wing reduced and joined up at the tip region at α = 10° by the virtue of occurring separation bubble and these vibrational modes became a chordwise mode of two at α = 12° as seen in **Figure 22**. The holes

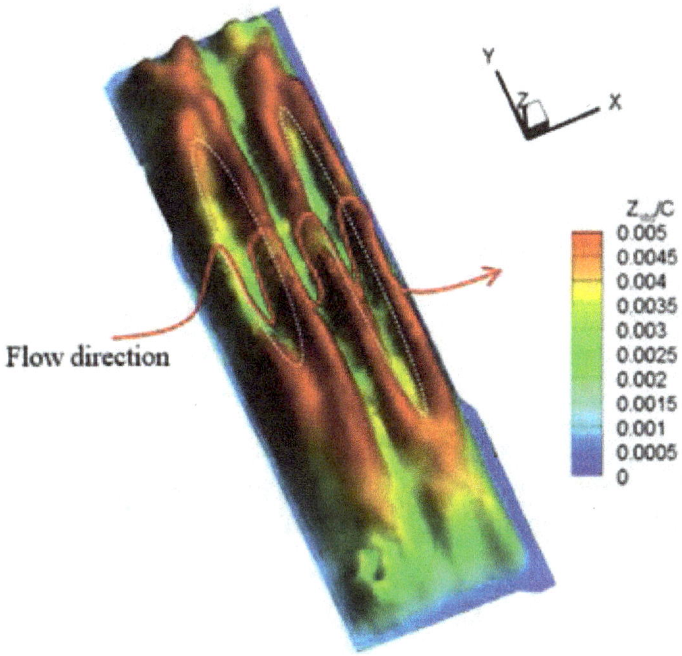

Figure 21. Three-dimensional view of standard deviation of mean deformation of AR = 3 flexible membrane wing at $\alpha = 10°$ for $Re = 5 \times 10^4$.

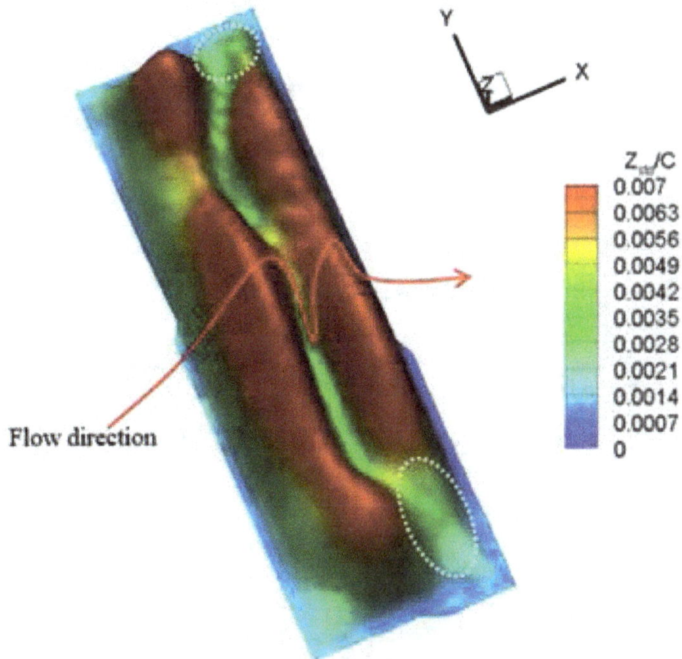

Figure 22. Three-dimensional view of standard deviation of mean deformation of AR = 3 flexible membrane wing at $\alpha = 12°$ for $Re = 5 \times 10^4$.

formed by the separation bubble in the middle of the wing were illustrated with white dashed lines and the regions with red color showed the peaks.

3.12. Partially flexible airfoil

The last major control device, which is the objective of the chapter, among passive flow controllers is the flexible membrane used on the surface of the airfoil. This type of airfoil is called as a segmented or partially flexible airfoil. Since it is a new concept of flow control method, a detailed investigation of a partially flexible membrane is rarely studied in the aerodynamic literature. A pioneered computational fluid dynamics (CFD) analysis was performed using

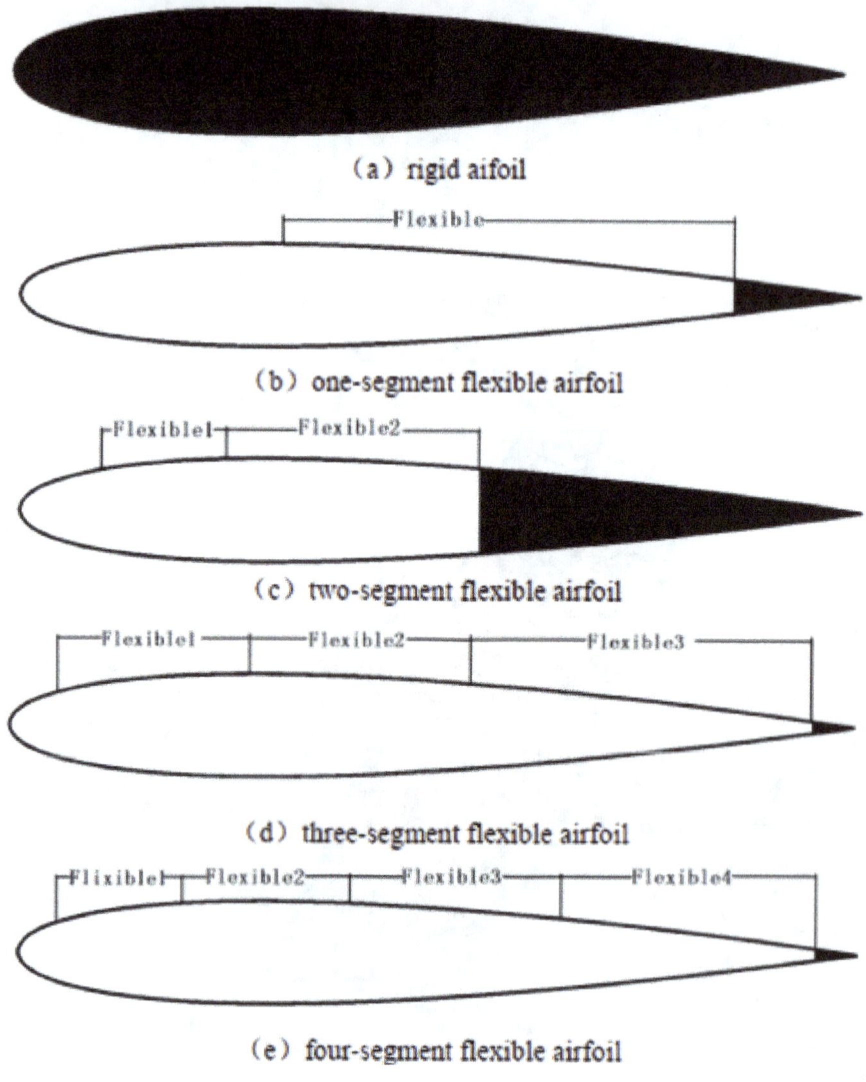

Figure 23. Different types of segmented airfoils [95].

flexible membrane material on the airfoil surface by using ANSYS software [95]. The fluid-structure interaction (FSI) method was used for numerical modeling to investigate interactions between fluid and membrane. The segmented airfoil is seen in **Figure 23** [95]. The flexible

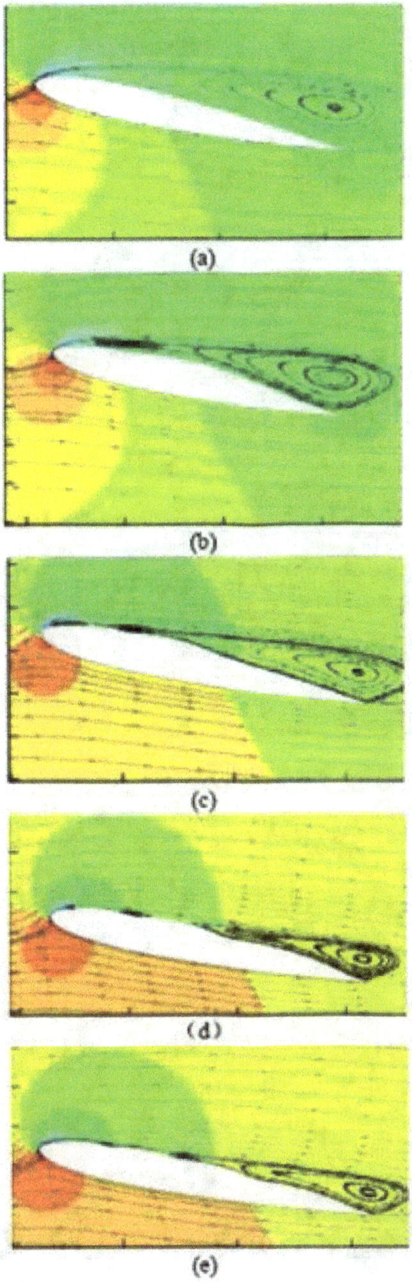

Figure 24. Streamline of rigid and flexible airfoils, $\alpha = 13°$: (a) rigid; (b) one-segment; (c) two-segment; (d) three-segment; and (e) four-segment [95].

Figure 25. Configuration of the partially flexible airfoil [99].

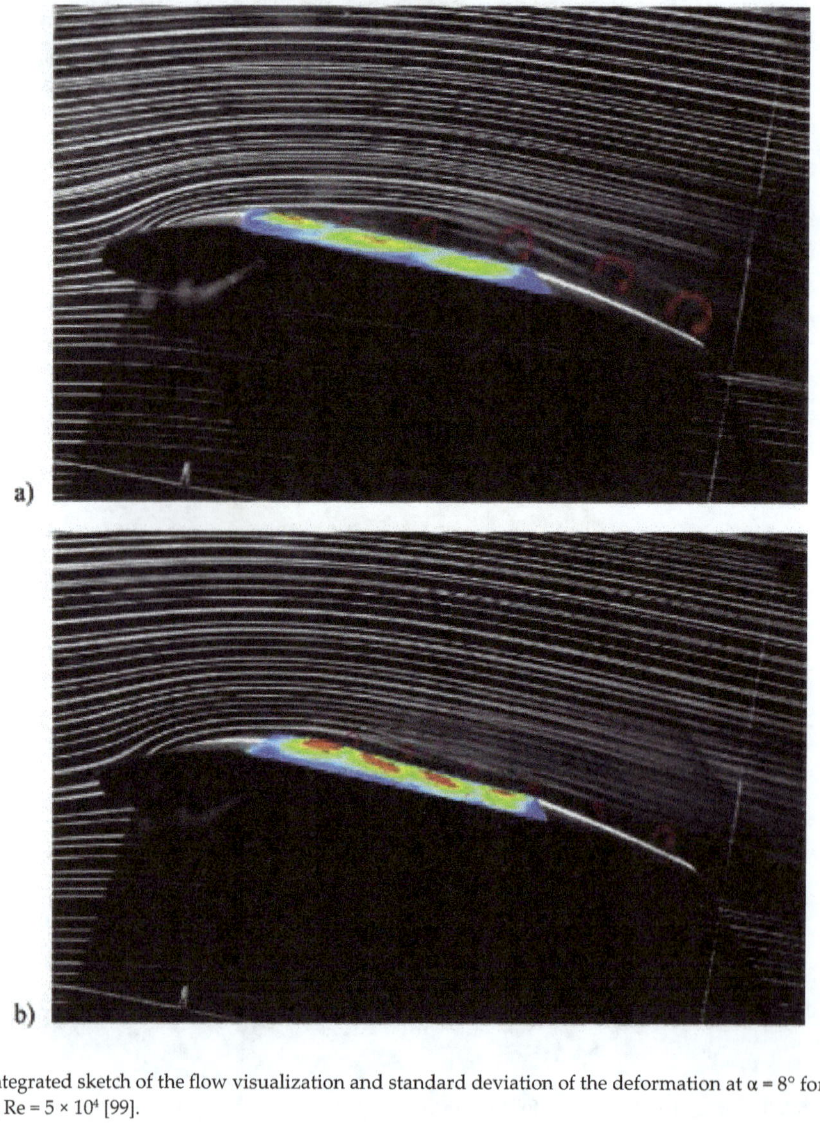

Figure 26. Integrated sketch of the flow visualization and standard deviation of the deformation at $\alpha = 8°$ for (a) $Re = 2.5 \times 10^4$ and (b) $Re = 5 \times 10^4$ [99].

membrane material was used on the suction side of the airfoil. They numerically modeled four different cases on which the upper surface of the airfoil was flexible. In this numerical model, the effect of flexibility on aerodynamic performance in various regions on the airfoil was investigated for a Reynolds number of 1.35×10^5. **Figure 24** [95] gives information about flow over the uncontrolled airfoil and the segmented flexible airfoils. It has been observed that the interaction between flow and the segmented airfoil decreases flow separations at high angles of attack. It has been found that the airfoil with three separate flexible zones shows the best aerodynamic performance and increased the lift coefficient by 39% compared to the rigid airfoil around the stall angle.

Apart from numerical study, first detailed experimental investigations on a partially flexible airfoil at low Reynolds numbers were carried out by Açıkel and Genç [99]. They modified the rigid NACA 4412 airfoil by using a membrane material that was located on the upper side of the airfoil as denoted in **Figure 25** [99]. The location of the membrane was between $x/c = 0.2$ and $x/c = 0.7$. In this study, different experimental methods such as force measurement, velocity measurement, deformation measurement, and smoke wire visualization were used to investigate flow control on partially flexible membrane airfoil.

According to the experimental results, flow control with flexibility is more effective at low angles of attack. **Figure 26** [99] demonstrates a combined sketch of the membrane standard deviation and smoke wire visualization for $\alpha = 8°$. This sketch showed that the membrane vibration modes were increased with increasing Reynolds number.

4. Conclusion

A detailed review with regard to passive control methods affecting the flow especially at low Reynolds numbers was presented in this chapter. The main purpose of this study is to clarify the passive control techniques for UAVs and MAVs operating at low Reynolds numbers. Besides the explanation of those techniques, especially three passive flow methods at low Reynolds numbers have been highlighted with their results as follows:

- Using the sandpaper as a passive flow controller [8, 69] on the surface of the airfoil has caused the LSB's size to reduce enormously, resulting in aerodynamic performance recovery. Moreover, the roughness-induced transition phenomenon also mentioned in studies performed by Puckert and Rist [100] and Bucci et al. [101] has occurred with the usage of sandpaper. Therefore, a more stable flow characteristic has been obtained.
- Time-dependent attitudes of LSB obtained from results of flexible membrane wings [74] showed that bubble size first increased and then reduced at different low Reynolds numbers and angles of attack. That is, it can be understood that fluid-structure interaction positively exhibited a good effect on aerodynamic performance by varying the bubble size with time.
- Regarding the usage of the partially flexible airfoil [99], flow control in conjunction with partial flexibility is more effective especially at low angles of attack. Both flow and flexibility-induced undulation over membrane material have caused the vibration modes, helping

the bubble size to be reduced. Thus, better aerodynamic performance with the increasing of lift coefficient has been obtained.

Consequently, this detailed chapter will present a comprehensive, practical, effective roadmap for the aerodynamic researchers especially interested in the flow control techniques over wind turbine blade or MAV applications operating at low Reynolds number regimes.

Nomenclature

Abbreviation

LSB	laminar separation bubble
UAV	unmanned aerial vehicle
MAV	micro air vehicle
Ma	Mach number
2D	two dimensional
3D	three dimensional
POD	proper orthogonal decomposition
DIC	digital image correlation
AR	aspect ratio
FSI	fluid-structure interaction
LE	leading edge
TE	trailing edge
VG	vortex generator
CFD	computational fluid dynamics
APG	adverse pressure gradient
AoA	angle of attack

Symbols

Re	Reynolds number
Re_c	critical Reynolds number
c	chord length of airfoil

s	span length of membrane wing
h	height of VG
b	cropped edge length of VG
L	vane length of VG
D	long gaps among vanes
d	short gaps among vanes
β	inflow angle
X	slot position
γ	slot width
ψ	angle between slot axis and chord normal
t	time
C_L	lift coefficient
C_D	drag coefficient
k	roughness height
α	angle of attack
$C_{L, max}$	maximum lift coefficient

Subscripts

L	lift
D	drag
max	maximum
c	critical

Author details

Mustafa Serdar Genç[1]*, Kemal Koca[1], Hacımurat Demir[1,2] and Halil Hakan Açıkel[1]

*Address all correspondence to: musgenc@erciyes.edu.tr

1 Wind Engineering and Aerodynamic Research Center, Department of Energy Systems Engineering, Erciyes University, Kayseri, Turkey

2 Department of Mechanical Engineering, Aksaray University, Aksaray, Turkey

References

[1] Arcara PC Jr, Bartlett DW, McCullers LA. Analysis for the application of hybrid laminar flow control to a long-range subsonic transport aircraft. Society of Automotive Engineers SAE 9121. 1991

[2] Horton HP. Laminar separation in two and three-dimensional incompressible flow [PhD thesis]. University of London; 1968

[3] Gaster M. The structure and behaviour of laminar separation bubbles. ARC Reports and Memoranda 3595. London, UK: Aeronautical Research Council (ARC); 1967

[4] Genç MS, Karasu I, Açıkel HH, Akpolat MT. Low Reynolds number flows and transition, low Reynolds number aerodynamics and transition. In: Serdar GENÇ M, editor. Rijeka, Croatia: InTech-Open Access; 2012. pp. 1-28. ISBN: 978-953-51-0492-6

[5] Diwan SS, Chetan SJ, Ramesh ON. On the bursting criterion for laminar separation bubbles. In: IUTAM Symposium on Laminar-Turbulent Transition. Dordrecht: Springer; 2006. pp. 401-407

[6] Boiko AV, Grek GR, Dovgal AV, Kozlov VV. The Origin of Turbulence in near-Wall Flows. Berlin, Germany: Springer Science & Business Media; 2013

[7] Dovgal AV, Kozlov VV, Michalke A. Laminar boundary layer separation: Instability and associated phenomena. Progress in Aerospace Sciences. 1994;**30**(1):61-94

[8] Genç MS, Koca K, Açikel HH. Investigation of pre-stall flow control on wind turbine blade airfoil using roughness element. Energy. 2019;**176**:320-334

[9] Prandtl L. Bemerkungen über die Entstehung der Turbulenz. ZAMM-Journal of Applied Mathematics and Mechanics/Zeitschrift für Angewandte Mathematik und Mechanik. 1921;**1**(6):431-436

[10] Tollmien W. Über die entstehung der turbulenz. In: Vorträge aus dem Gebiete der Aerodynamik und verwandter Gebiete. Berlin, Heidelberg: Springer; 1930. pp. 18-21

[11] Schubauer GB, Skramstad HF. Laminar boundary layer oscillations and the stability of laminar flow. Journal of Aerosol Science. 1947;**14**:69-78

[12] Mayle RE. The role of laminar-turbulent transition in gas turbine engines. Journal of Turbomachinery. 1991;**113**:509-537

[13] Kurelek J. Transition in a laminar separation bubble and the effect of acoustic excitation [master's thesis]. University of Waterloo; 2016

[14] Bertolotti F. Linear and nonlinear stability of boundary layers with streamwise varying properties [Ph.D. thesis]. 1991

[15] Schlichting H, Gersten K. Boundary Layer Theory. Berlin, Heidelberg: Springer-Verlag; 2000

[16] Langtry R, Menter F. Overview of Industrial Transition Modelling in CFX. Otterfing, Germany: ANSYS Germany GmbH, ANSYS CFX; 2006

[17] Narasimha R, Sreenivasan KR. Relaminarization of fluid flows. Advances in Applied Mechanics. vol. 19. 1979. pp. 221-309

[18] Badrya C, Govindarajan B, Chopra I. Basic understanding of unsteady airfoil aerodynamics at low Reynolds numbers. In: 2018 AIAA Aerospace Sciences Meeting. 2018. p. 2061

[19] Pechlivanoglou G. Passive and Active Flow Control Solutions for Wind Turbine Blades. 2013

[20] Seshagiri A, Cooper E, Traub LW. Effects of vortex generators on an airfoil at low Reynolds numbers. Journal of Aircraft. 2009;**46**(1):116-122

[21] Aider JL, Beaudoin JF, Wesfreid JE. Drag and lift reduction of a 3D bluff-body using active vortex generators. Experiments in Fluids. 2010;**48**(5):771-789

[22] Lee S, Loth E. Supersonic boundary-layer interactions with various micro-vortex generator geometries. The Aeronautical Journal. 2009;**113**(1149):683-697

[23] Wang H, Zhang B, Qiu Q, Xu X. Flow control on the NREL S809 wind turbine airfoil using vortex generators. Energy. 2017;**118**:1210-1221

[24] Langan K, Samuels J. Experimental investigation of maneuver performance enhancements on an advanced fighter/attack aircraft. In: 33rd Aerospace Sciences Meeting and Exhibit. 1995. p. 442

[25] Farzaneh-Gord M, Sadi M. Improving vortex tube performance based on vortex generator design. Energy. 2014;**72**:492-500

[26] Zhou G, Pang M. Experimental investigations on thermal performance of phase change material—Trombe wall system enhanced by delta winglet vortex generators. Energy. 2015;**93**:758-769

[27] Taylor HD. The elimination of diffuser separation by vortex generators. Technical Report No, 4012, 3. East Hartford, CT: United Aircraft Corporation; 1947

[28] Schubauer GB, Spangenberg WG. Forced mixing in boundary layers. Journal of Fluid Mechanics. 1960;**8**(1):10-32

[29] Lin JC, Robinson SK, McGhee RJ, Valarezo WO. Separation control on high-lift airfoils via micro-vortex generators. Journal of Aircraft. 1994;**31**(6):1317-1323

[30] Nickerson JR. A study of vortex generators at low Reynolds numbers. In: 24th Aerospace Sciences Meeting. 1986. p. 155

[31] Gao L, Zhang H, Liu Y, Han S. Effects of vortex generators on a blunt trailing-edge airfoil for wind turbines. Renewable Energy. 2015;**769**:303-311

[32] Bragg MB, Gregorek GM. Experimental study of airfoil performance with vortex generators. Journal of Aircraft. 1987;**24**(5):305-309

[33] Baldacchino D, Ferreira C, Tavernier DD, Timmer WA, van Bussel GJW. Experimental parameter study for passive vortex generators on a 30% thick airfoil. Wind Energy. Vol. 21. 2018. pp. 745-765

[34] Godard G, Stanislas M. Control of a decelerating boundary layer. Part 1: Optimization of passive vortex generators. Aerospace Science and Technology. 2006;**10**(3):181-191

[35] Handley PF. Wing and similar member of aircraft. U.S. Patent No 1,427,012. 1922

[36] Van Dam CP. The aerodynamic design of multi-element high-lift systems for transport airplanes. Progress in Aerospace Sciences. 2002;**38**(2):101-144

[37] Richard PR, Wilkins SJ, Hall JW. Particle image velocimetry investigation of the coherent structures in a leading-edge slat flow. Journal of Fluids Engineering. 2018;**140**(4):041105

[38] Rumsey CL, Ying SX. Prediction of high lift: Review of present CFD capability. Progress in Aerospace Sciences. 2002;**38**(2):145-180

[39] Weick FE, Shortal JA. The Effect of Multiple Fixed Slots and a Trailing-Edge Flap on the Lift and Drag of a Clark y Airfoil. 1933

[40] Weick FE, Platt RC. Wind-Tunnel Tests on Model Wing with Fowler Flap and Specially Developed Leading-Edge Slot. 1933

[41] Krüger W. Wind-Tunnel Investigations on a Changed Mustang Profile with Nose Flap Force and Pressure-Distribution Measurements. 1947

[42] Genç MS, Kaynak Ü, Lock GD. Flow over an aerofoil without and with a leading-edge slat at a transitional Reynolds number. Proceedings of the Institution of Mechanical Engineers, Part G: Journal of Aerospace Engineering. 2009;**223**(3):217-231

[43] Fish FE, Lauder GV. Passive and active flow control by swimming fishes and mammals. Annual Review of Fluid Mechanics. 2006;**38**:193-224

[44] Ibrahim IH, New TH. Tubercle modifications in marine propeller blades. In: 10th Pacific Symposium on Flow Visualization and Image Processing. Italy: Naples. p. 2015

[45] Post ML, Decker R, Sapell AR, Hart JS. Effect of bio-inspired sinusoidal leading-edges on wings. Aerospace Science and Technology. 2018;**81**:128-140

[46] Wang Z, Zhuang M. Leading-edge serrations for performance improvement on a vertical-axis wind turbine at low tip-speed-ratios. Applied Energy. 2017;**208**:1184-1197

[47] Cai C, Zuo Z, Liu S, Maeda T. Effect of a single leading-edge protuberance on NACA 634-021 airfoil performance. Journal of Fluids Engineering. 2018;**140**(2):021108

[48] Wei Z, New TH, Cui YD. An experimental study on flow separation control of hydrofoils with leading-edge tubercles at low Reynolds number. Ocean Engineering. 2015;**108**:336-349

[49] Belamadi R, Djemili A, Ilinca A, Mdouki R. Aerodynamic performance analysis of slotted airfoils for application to wind turbine blades. Journal of Wind Engineering and Industrial Aerodynamics. 2016;**151**:79-99

[50] Beyhaghi S, Amano RS. A parametric study on leading-edge slots used on wind turbine airfoils at various angles of attack. Journal of Wind Engineering and Industrial Aerodynamics. 2018;**175**:43-52

[51] Shi X, Xu S, Ding L, Huang D. Passive flow control of a stalled airfoil using an oscillating micro-cylinder. Computers & Fluids. 2019;**178**:152-165

[52] Luo D, Huang D, Sun X. Passive flow control of a stalled airfoil using a microcylinder. Journal of Wind Engineering and Industrial Aerodynamics. 2017;**170**:256-273

[53] Wang Y, Li G, Shen S, Huang D, Zheng Z. Investigation on aerodynamic performance of horizontal axis wind turbine by setting micro-cylinder in front of the blade leading edge. Energy. 2018;**143**:1107-1124

[54] Wang JJ, Li YC, Choi KS. Gurney flap-lift enhancement, mechanisms and applications. Progress in Aerospace Sciences. 2008;**44**(1):22-47

[55] Liebeck RH. Design of subsonic airfoils for high lift. Journal of Aircraft. 1978;**15**(9):547-561

[56] Zhu B, Huang Y, Zhang Y. Energy harvesting properties of a flapping wing with an adaptive Gurney flap. Energy. 2018;**152**:119-128

[57] Shukla V, Kaviti AK. Performance evaluation of profile modifications on straight-bladed vertical axis wind turbine by energy and Spalart Allmaras models. Energy. 2017;**126**:766-795

[58] Meyer R, Hage W, Bechert DW, Schatz M, Knacke T, Thiele F. Separation control by self-activated movable flaps. AIAA Journal. 2007;**45**(1):191-199

[59] Rosti ME, Omidyeganeh M, Pinelli A. Passive control of the flow around unsteady aerofoils using a self-activated deployable flap. Journal of Turbulence. 2018;**19**(3):204-228

[60] Arivoli D, Singh I. Self-adaptive flaps on low aspect ratio wings at low Reynolds numbers. Aerospace Science and Technology. 2016;**59**:78-93

[61] Schluter JU. Lift enhancement at low Reynolds numbers using self-activated movable flaps. Journal of Aircraft. 2010;**47**(1):348-351

[62] Kasper W. Aircraft wing with vortex generation. U.S. Patent No 3,831,885. 1974

[63] Olsman WFJ, Colonius T. Numerical simulation of flow over an airfoil with a cavity. AIAA Journal. 2011;**49**(1):143-149

[64] Lam GC, Leung RC. Aeroacoustics of NACA 0018 airfoil with a cavity. AIAA Journal. 2018;**56**(3):1-12

[65] Li Q, Liu C. LES for supersonic ramp control flow using MVG at $M = 2.5$ and $Re_\theta = 1440$. In: 48th AIAA Aerospace Sciences Meeting Including the New Horizons Forum and Aerospace Exposition. 2010. p. 592

[66] Huebsch WW. Numerical Investigation on the Interaction between Surface Roughness and Viscous Flows. 2000

[67] Koca K, Genç MS, Açıkel HH, Çağdaş M, Bodur TM. Identification of flow phenomena over NACA 4412 wind turbine airfoil at low Reynolds numbers and role of laminar separation bubble on flow evolution. Energy. 2018;**144**:750-764

[68] Genç MS, Koca K, Açıkel HH, Özkan G, Kırış MS, Yıldız R. Flow characteristics over NACA4412 airfoil at low Reynolds number. In: EPJ Web of Conferences. EDP Sciences. 2016. p. 02029

[69] Koca K, Genç MS, Açikel HH. Rüzgar Türbini Kanadı Üzerindeki Yüzey Pürüzlülüğü Etkisinin Deneysel İncelenmesi. Çukurova Üniversitesi Mühendislik-Mimarlık Fakültesi Dergisi. 2016;**31**:127-134

[70] Shyy W, Berg M, Ljungqvist D. Flapping and flexible wings for biological and micro air vehicles. Progress in Aerospace Sciences. 1999;**35**(5):455-505

[71] Timpe A, Zhang Z, Hubner J, Ukeiley L. Passive flow control by membrane wings for aerodynamic benefit. Experiments in Fluids. 2013;**54**(3):1471

[72] Smith MJ. Simulating moth wing aerodynamics-towards the development of flapping-wing technology. AIAA Journal. 1996;**34**(7):1348-1355

[73] Mueller TJ, DeLaurier JD. Aerodynamics of small vehicles. Annual Review of Fluid Mechanics. 2003;**35**(1):89-111

[74] Demir H, Genç MS. An experimental investigation of laminar separation bubble formation on flexible membrane wing. European Journal of Mechanics-B/Fluids. 2017;**65**:326-338

[75] Genç MS, Demir H, Açıkel HH. Time dependent laminar separation bubble formation and deformation over flexible membrane wing. In: The International Symposium on Sustainable Aviation (ISSA-2016); 29 May-1 June 2016; Istanbul, Turkey

[76] Genç MS, Açikel HH, Demir H, Özden M, Çağdaş M, Isabekov I. Effect of tip vortices on membrane vibration of flexible wings with different aspect ratios. In: EPJ Web of Conferences. EDP Sciences. 2016. p. 02028

[77] Genç MS, Özden M, Açikel HH, Demir H, Isabekov I. Unsteady flow over flexible wings at different low Reynolds numbers. In: EPJ Web of Conferences. EDP Sciences. 2016. p. 02030

[78] Rojratsirikul P, Genc MS, Wang Z, Gursul I. Flow-induced vibrations of low aspect ratio rectangular membrane wings. Journal of Fluids and Structures. 2011;**27**(8):1296-1309

[79] Rojratsirikul P, Wang Z, Gursul I. Effect of pre-strain and excess length on unsteady fluid-structure interactions of membrane airfoils. Journal of Fluids and Structures. 2010;**26**(3):359-376

[80] Genç MS. Unsteady aerodynamics and flow-induced vibrations of a low aspect ratio rectangular membrane wing with excess length. Experimental Thermal and Fluid Science. 2013;**44**:749-759

[81] Greenhalgh S, Curtiss HC, Smith B. Aerodynamic properties of a two-dimensional inextensible flexible airfoil. AIAA Journal. 1984;**22**(7):865-870

[82] Arbós-Torrent S, Ganapathi-subramani B, Palacios R. Leading-and trailing-edge effects on the aeromechanics of membrane aerofoils. Journal of Fluids and Structures. 2013; **38**:107-126

[83] Galvao R, Israeli E, Song A, Tian X, Bishop K, Swartz S, et al. The aerodynamics of compliant membrane wings modeled on mammalian flight mechanics. In: 36th AIAA Fluid Dynamics Conference and Exhibit. 2006. p. 2866

[84] Bleischwitz R, De Kat R, Ganapathisubramani B. Aeromechanics of membrane wings in ground-effect. In: 45th AIAA Fluid Dynamics Conference. 2015. p. 2764

[85] Hu H, Kumar AG, Abate G, Albertani R. An experimental investigation on the aerodynamic performances of flexible membrane wings in flapping flight. Aerospace Science and Technology. 2010;**14**(8):575-586

[86] Albertani R, Stanford B, Hubner JP, Ifju PG. Aerodynamic coefficients and deformation measurements on flexible micro air vehicle wings. Experimental Mechanics. 2007;**47**(5):625-635

[87] Osterberg N. Experimental investigation of dynamic stall on pliant wings for micro air vehicles. In: 54th AIAA Aerospace Sciences Meeting. 2016. p. 0146

[88] Song A, Tian X, Israeli E, Galvao R, Bishop K, Swartz S, et al. Aeromechanics of membrane wings with implications for animal flight. AIAA Journal. 2008;**46**(8):2096-2106

[89] Rojratsirikul P, Wang Z, Gursul I. Unsteady aerodynamics of membrane airfoils. In: 46th AIAA Aerospace Sciences Meeting and Exhibit. 2008. p. 613

[90] Rojratsirikul P, Wang Z, Gursul I. Unsteady fluid–structure interactions of membrane airfoils at low Reynolds numbers. Experiments in Fluids. 2009;**46**(5):859

[91] Tamai M, Murphy J, Hu H. An experimental study of flexible membrane airfoils at low Reynolds numbers. In: 46th AIAA Aerospace Sciences Meeting and Exhibit. 2008. p. 580

[92] Wrist AH, Zhang Z, Pepley D, Hubner JP. Aerodynamic comparison of flat and cambered frames for flexible MAV wings. In: 53rd AIAA Aerospace Sciences Meeting. 2015. p. 1299

[93] Attar PJ, Morris BJ, Romberg WA, Johnston JW, Parthasarathy RN. Experimental characterization of aerodynamic behavior of membrane wings in low-Reynolds-number flow. AIAA Journal. 2012;**50**(7):1525-1537

[94] Viieru D, Tang J, Lian Y, Liu H, Shyy W. Flapping and flexible wing aerodynamics of low Reynolds number flight vehicles. In: 44th AIAA Aerospace Sciences Meeting and Exhibit. 2006. p. 503

[95] Hefeng D, Chenxi W, Shaobin L, Zhen SX. Numerical research on segmented flexible airfoils considering fluid-structure interaction. Procedia Engineering. 2015;**99**:57-66

[96] Lian Y, Shyy W. Laminar-turbulent transition of a low Reynolds number rigid or flexible airfoil. AIAA Journal. 2007;**45**(7):1501-1513

[97] Gordnier RE, Attar PJ. Impact of flexibility on the aerodynamics of an aspect ratio two membrane wing. In: ASME 2012 Fluids Engineering Division Summer Meeting collocated with the ASME 2012 Heat Transfer Summer Conference and the ASME 2012 10th International Conference on Nanochannels, Microchannels, and Minichannels. American Society of Mechanical Engineers; 2012. pp. 1481-1496

[98] Demir H. Investigation of Unsteady Aerodynamics of Flexible Wings at Low Reynolds Numbers, PhD. Thesis. Turkey: Erciyes University; 2018

[99] Açıkel HH, Genç MS. Control of laminar separation bubble over wind turbine airfoil using partial flexibility on suction surface. Energy. 2018;**165**:176-190

[100] Puckert DK, Rist U. Experiments on critical Reynolds number and global instability in roughness-induced laminar-turbulent transition. Journal of Fluid Mechanics. 2018;**844**:878-904

[101] Bucci MA, Puckert DK, Andriano C, Loiseau JC, Cherubini S, Robinet JC, et al. Roughness-induced transition by quasi-resonance of a varicose global mode. Journal of Fluid Mechanics. 2018;**836**:167-191

5

Design and Estimation of an AUV Portable Intelligent Rescue System Based on Attitude Recognition Algorithm

Sheng-Chih Shen and Yi-Ting Yang

Abstract

This research is based on the attitude sensing algorithm to design a portable intelligent rescue system for autonomous underwater vehicles (AUVs). To lower the possibility of losing the underwater vehicle and reduce the difficulty of rescuing, when an AUV intelligent rescue system (AIRS) detects the fault of AUVs and they could not be reclaimed, AIRS can pump carbon dioxide into the airbag immediately to make the vehicle resurface. AIRS consists of attitude sensing module, double-trigger inflator mechanism, and activity recognition algorithm. The sensing module is an eleven-DOF sensor that is made up of a six-axis inertial sensor, a three-axis magnetometer, a barometer, and a thermometer. Furthermore, the signal calibration and extended Kalman filter (SC-EKF) is proposed to be used subsequently to calibrate and fuse the data from the sensing module. Then, the attitude data are classified with the principle of feature extraction (FE) and backpropagation network (BPN) classifier. Finally, the designed double-trigger inflator can be triggered not only by electricity but also by water damage when the waterproof cabin is severely broken. With the AIRS technology, the safety of detecting and investigating the use AUVs can be increased since there is no need to send divers to engage in the rescue mission under water.

Keywords: AUVs, sensing module, intelligent rescue system, extended kalman filter, feature extracted, back propagation neural network

1. Introduction

Significant research is being conducted on the development of autonomy for underwater robotic vehicles, which are widely employed in many fields of application such as oceanographic, marine archeology [1], military organizations, and cable tracking and inspection [2].

The advent of underwater robotic vehicles has significantly reduced the dangers in deep sea exploration. Two kinds of robotic vehicles used in marine research are remotely operated vehicles (ROVs) and autonomous underwater vehicles (AUVs). The main difference between the two is that ROVs are connected to the ship by communication cables whereas AUVs operate independently from the ship. AUVs operate without an umbilical; therefore, AUVs are able to conduct a larger range of work area. In this research, AUV intelligent rescue system (AIRS) can minimize the loss of losing or unpredictable catastrophic failure of AUVs caused by emergency conditions.

In recent years, the micro-electro-mechanical system (MEMS) has almost become a vital technology for modern society because of its small volume, low power consumption, low cost, and ease of integration into systems or modification [3]. The MEMS technology creates entirely innovative kinds of products, such as gyroscope sensor in camera-shake detection systems [4], multi-axis inertial motion sensors for smartphone-based navigation [5], and rehabilitation systems based on inertial measurement units (IMUs) [6]. Additionally, the integration of global positioning system and inertial navigation system (GPS/INS) is usually employed to measure the position of AUVs [7, 8]. Unfortunately, small errors in the measurement of initial data are double integrated into larger errors progressively in attitude data, and such errors increase without bound. Error reduction calibration for initialization of INS is paramount for the systematic parameter, like scale factor, bias, and misalignment of the axes. As the situation depicted above, a calibration method is investigated and adopted in this research. There are many researches utilizing Kalman filter (KF) [9], complementary filter (CF) [10], adaptive Kalman filter (AKF) [11], or extended Kalman filter (EKF) [12] to fuse the gyroscope and accelerometer together, taking advantages of their individual strength. In this research, we use EKF to filter IMU outputs with a balance of noise canceling and adaptability simultaneously used in sensing attitude algorithm for AIRS.

The sensing algorithm with EKF for sensing the attitude of AUV, while employing the backpropagation network (BPN) to classify motion data that are formed in the AUV. Before classify the data, there is a problem. The problem is compounded by the fact that our system is computationally complex due to values estimation of AUV and causes the high dimension that is called "curse of dimensionality," which donates the drastic rise of computational complexity and the classification error. Therefore, the step of feature extraction (FE) is an important procedure in data classification [26]. Further, we feed the data into eural network to distinguish every AUV motion in each operating period. The neural network consists of multiple artificial neurons to receive inputs, and process them to obtain an output [13]. By repeating amendments to the model weights, the neural network makes central processing unit (CPU) more logical to calculate the nonlinear systems of AUV motion [14]. The neural network has been widely used because of a number of advantages, including estimating which variables are important in classification, detecting possible interactions between predictor variables, and constructing the prediction model with a high accuracy rate. Furthermore, the neural network has emerged as an important tool in classification which has been investigated in many different important applications [15]. In this study, neural network classifier (NNC) is used in the calculation of nonlinear systems AUV attitude data to detect a catastrophic failure of an AUV. Based on the multilayer feed-forward backpropagation algorithm as NNC, we proposed an effective activity recognition method using 10-DOF sensor module.

The concept of AIRS is inspired from the vehicle safety device: airbags have a flexible fabric bag that can protect and restrict the occupant from being thrown away during a crash accident [16, 17]. Wang and Dragcevic designed the airbags, which fire via a small pyrotechnic charge to increase motorcyclists' safety protections when riding a motorcycle [18, 19]. The airbags are applied to not only fall-protection devices but also automatic inflatable life buoy by the inflatable method of a gas cylinder [20–23]. We extended the airbag systems to load on the AUV with CO_2 cylinder to implement the AIRS. In this research, we developed a malformed detection algorithm for AUV with both EKF and BPN, and the implementation of the airbag system loaded on AUV is underway.

2. Intelligent rescue system

In this section, we introduce the hardware components of an AIRS and its working principles. The hardware of AIRS contains the sensing module and a double-trigger inflator. The 11-DOF sensing module combines the 9-DOF inertial sensor, barometric pressure sensor, and temperature sensor to collect the motion data and the depth of the underwater vehicle. The inflator is used with gas cylinders as the double trigger mechanism, which can be triggered not only by electricity but also by water damage when the waterproof cabin is severely broken. Furthermore, we used the commercial software LabVIEW™ to analyze the relation between buoyancy force and inflator device in water, to confirm the inflator specifications (**Figure 1**).

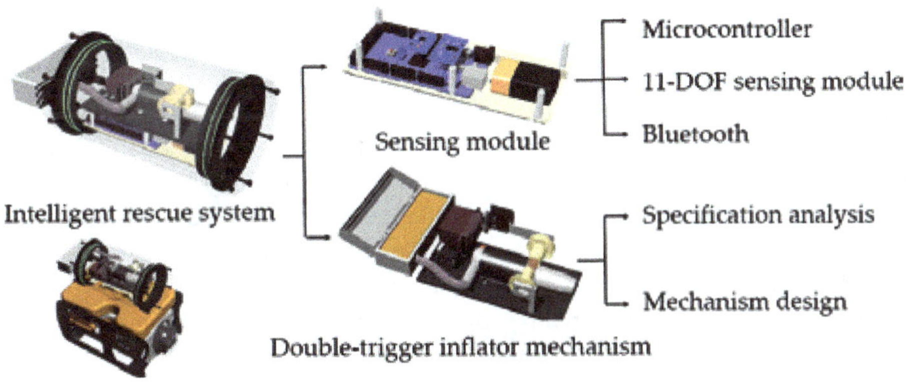

Figure 1. The structure of the intelligent rescue system.

2.1. Sensing module

Master control module uses ATmega2560 with the Arduino program written to read motion data of sensor. Ten-axis motion sensor combined with the accelerometer, gyroscope, magnetometer, and the barometer to acquire attitude angle data through the inter-integrated circuit (I2C) is adopted. The communication module uses Bluetooth transmit data wirelessly to the processing software. The block diagram of the 11-DOF sensing module is shown in **Figure 2**.

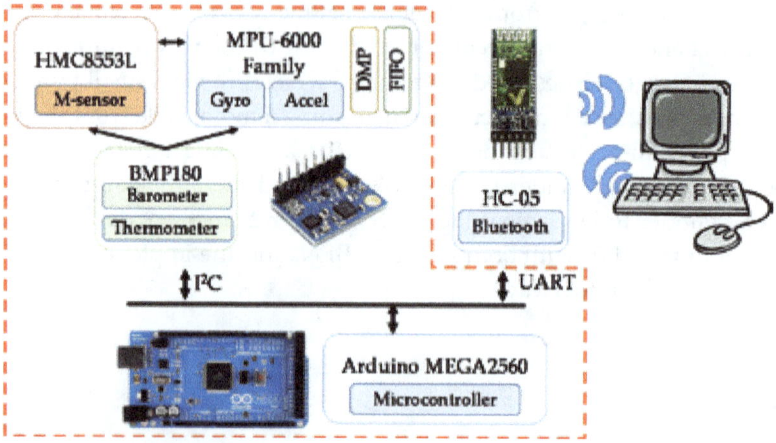

Figure 2. The block diagram of the sensing module.

2.2. Inflator mechanism

In case of failure and sinking of AUVs underwater, we designed an airbag system to increase the buoyancy and reduce the overall density of the AUVs. AIRS is installed on top of AUV, an airbag inflated in a fault condition to force the AUVs to surface up. In this section, we are going to introduce how to manufacture an inflator mechanism of the AIRS for the AUV, including the specification analysis and the mechanism design.

2.2.1. Specification analysis

The AIRS mechanism is installed on top of the AUV and dives down into the ocean. One should note that the hydrostatic pressure exerted by a liquid increases with increase in ocean depth, which influences the specification of the cylinder. Not only that, the full displacement and weight of the AUV also affect the size of the airbag and cylinder. These interrelated factors cause the different performance of the airbag inflation time and rescue time. To analyze the inflator specification more accurately and conveniently, we have engineered the specification calculator for the inflator to ensure that the airbag can be inflated and can bring the AUV to surface successfully. From the required buoyancy and the depth of the location, the calculator analyzes the required gas cylinders and airbag volume.

From the result of the calculation illustrated in **Figure 3(b)**, it takes 68.1 seconds to make nylon airbags fill with CO2 and provides 42.1 N buoyancy for AUV. The design of added buoyancy can bring 7.47 kgw of underwater vehicle to the surface (assume that the full load displacement is fixed). With this technology, we can install the appropriate number of gas cylinders and airbags in accordance with different displacements of AUV accurately and conveniently.

2.2.2. Mechanism design

After the above analysis by the calculator, we found out the appropriate gas cylinder and airbag, and designed the corresponding mechanism parts for the selected gas cylinder and airbag. We are going to elaborate on the design concept of the inflator mechanism in the following. **Figure 4** illustrates a double-trigger inflator including a servomotor, horn, spring,

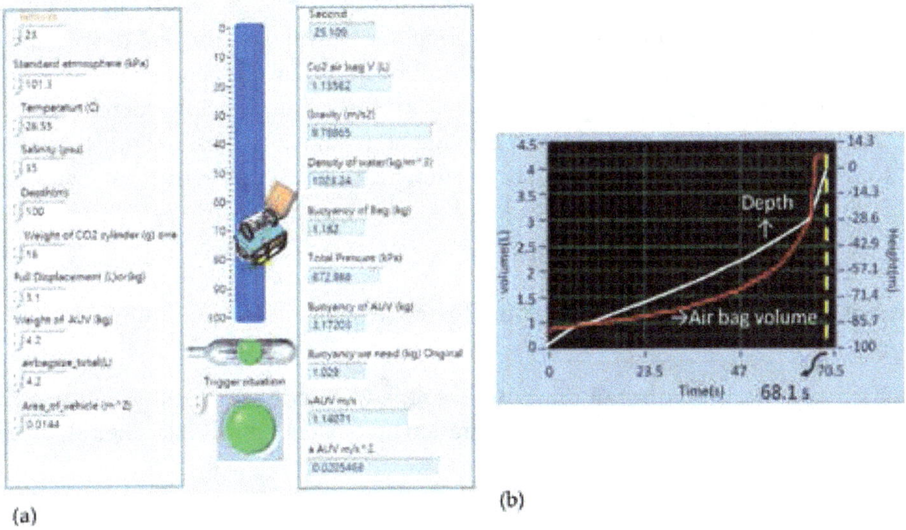

Figure 3. The result of calculation specification analysis. (a) The front panel of the specification calculator. (b) The growth rate between the water depth and the airbag volume.

spring case, crowbar-striker, water-soluble PVA fiber, release button, and CO_2 cylinder. The designed double-trigger inflator can be triggered not only by electricity but also by water damage when the waterproof cabin is severely broken.

After the above analysis by the calculator, we found out the appropriate gas cylinder and airbag, and design the corresponding mechanism parts for the selected gas cylinder and airbag.

There are two operating mechanisms to trigger the inflator. To begin with the electricity trigger mechanism as shown in **Figure 4(b)**, we assume that the waterproof cabin is watertight; after detecting the failure of AUVs, servo motor rotates for pushing the crowbar-striker, and the crowbar-striker punches the release button, which needs 35 N normal force to release into the airbag through the air tube. We chose MG995 for servomotor, which has product size 40.7 × 19.7 × 42.9 mm and weight 55 g; MG995 is sufficient to move the horn for the release button with its maximum torque 127 N-cm.

Besides, if the AUV is severely impacted and causes cabin seepage, so that the power system is damaged, the AUV cannot resurface by the thruster or the electricity-trigger inflator. Hence, we designed the water-trigger inflator for this kind of case. When the cabin is broken, water

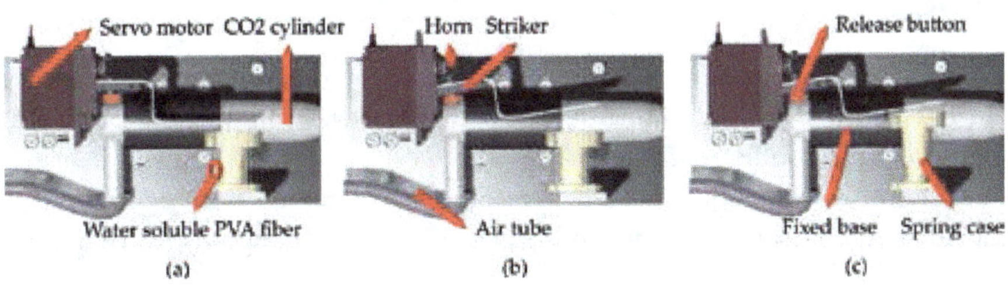

Figure 4. The design diagram of a double-trigger inflator. (a) before release. (b) release by servo motor. (c) release by spring.

melts the water-soluble PVA fiber and releases the spring case as shown in **Figure 4(c)**. Within the spring case, the music wire is selected to spring; the true maximum load of spring is 27.985 N, which is not enough to press the release button. Thus, we extended the striker into a crowbar-striker, and amplified the spring force by the lever principle to release the push button. The final part is the airbag, which is made of nylon 6/6 and folded under a cover. The airbag weighs 150 g and the value of tensile strength is 809.3 atm, which can withstand the pressure caused by gas cylinders.

3. Sensor with activity recognition algorithm

The method and the procedure of data processing and activity recognition are described in Section 3. The former includes the signal calibration and extended Kalman filter algorithm; the latter is constructed of FE and BPN classifier.

3.1. Data processing of 11-DOF sensing module

The 10-DOF sensing module including the 9-DOF IMU (three accelerometers, three gyroscopes, three magnetometers), barometer, and thermometer integrated with a microcontroller (Arduino Mega2560). The six-axis inertial sensor (MPU6050), which is a complete triple-axis gyroscope and triple-axis accelerometer inertial sensing system, is the most suitable sensor for stabilization and attitude measurement. MPU6050 contains the digital motion processor (DMP) which performs the motion processing algorithm itself. However, the horizontal attitude (yaw) is not a possible measurement. We used a magnetometer (HMC5883L) with fusion algorithm accordingly for eliminating the gyroscope offset to recognize the AUV activity context more reliably. The barometer (BMP180) is used in AIRS to detect the depth of the AUV and control the pressure of the airbag for supplying the perfect buoyancy. In addition, BMP180 also includes a temperature sensor, which does not only measure the underwater temperature but also thermal compensation for the MPU6050. Although the development of the MEMS technology has made a great progress, the IMUs are still difficult to collect precise data in the presence of various errors. Calibration is the process of comparing the measurement outputs with known reference data and determining the coefficients that drive the output to agree with the reference data over a range of output. We calibrated the sensor in terms of its thermal noise, bias, and scale factor.

3.2. EKF strategy

A quaternion-based EKF is proposed in this section for determining the attitude of the AUV from the outputs of IMU. Attitude estimation is a very important part of the AIRS. If the initial data are double integrated into larger errors without bound, it brings on the misclassification of ANN algorithm with wrong attitude values. For the sake of this error, we used a series of measurements and observed over time for signal processing. The main

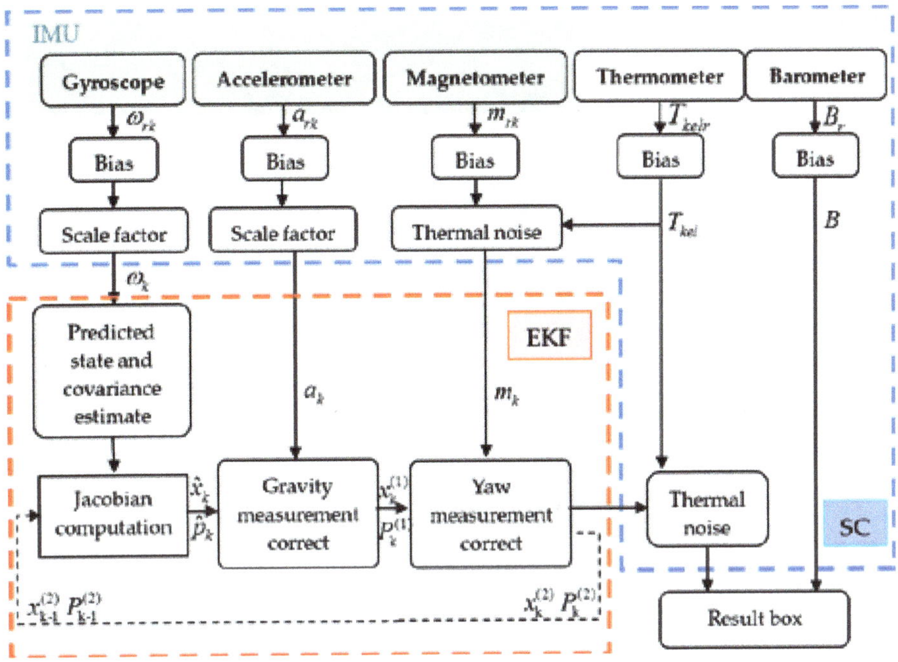

Figure 5. Overview of the SCEKF structure.

advantage of the KF is its ability to provide the quality of the estimate, but the KF only applies to linear and Gaussian models. The EKF conversely is a nonlinear version of the KF which linearizes about an estimate of the current mean and covariance. In view of this, we chose the EKF to filter IMU outputs with a balance of noise canceling and adaptability simultaneously, used in sensing attitude algorithm for AIRS. We proposed the EKF fusing with the accelerometer, gyroscope, and magnetometer integrated with sensor calibration (SC). SCEKF results in an improvement of the orientation accuracy from IMU. A flow-chart of structure performed by the proposed SCEKF is capsuled in **Figure 5**.

3.2.1. Prediction step

$$\hat{x}_k = \Phi(T_s, \hat{\omega}_{k-1})x_{k-1}^{(2)} + \hat{v}_{k-1} \tag{1}$$

$$\hat{P}_k = F_{k-1}P_{k-1}^{(2)}F_{k-1}{}^T + B_{k-1}QB_{k-1}^T \tag{2}$$

The first step of EKF predicts a current state and covariance matrix at time k. We estimated a current state based on the previous states $\hat{x}_{k-1} = [\hat{a}_{k-1}\hat{\omega}_{k-1}\hat{m}_{k-1}]^T$ that are composed of the gyroscope measurement $\hat{\omega}_{k-1}$ and white noise $\hat{v}_{k-1} = \left[{}^a\hat{v}_{k-1}{}^\omega\hat{v}_{k-1}{}^m\hat{v}_{k-1}\right]^T$ and the priori covariance matrix $\hat{P}_k$ based on a previous covariance matrix $P_{k-1}^{(2)}$, covariance Q and a state noise coefficient matrix B_{k-1}.

3.2.2. Jacobian computation

$$F_{k-1} = \frac{\partial \Phi}{\partial x}\Big|_{T_s, \hat{\omega}_{k-1}} = J_f(T_s, \hat{\omega}_{k-1}) \tag{3}$$

$$\tilde{y}_k = z_k - h(\hat{x}_k, 0), \quad H_k = \frac{\partial h}{\partial x}|_{\hat{x}_k} = J_h(\hat{x}_k) \tag{4}$$

However, in view of the nonlinear process of state transition F_{k-1} and observation H_k directly, the EKF approach requires estimation by computing the Jacobian.

3.2.3. Gravity corrections

$$S_k = H_k \hat{P}_k H_k^{\ T} + R_a \tag{5}$$

$$K_k = \hat{P}_k H_k^{\ T} \left(H_k \hat{P}_k H_k^{\ T} + R_a \right)^{-1} \tag{6}$$

$$x_k^{(1)} = \hat{x}_k + K_k(a_k - H_k \hat{x}_k) \tag{7}$$

$$P_k^{(1)} = (I_{4\times4} - K_k H_k)\hat{P}_k \tag{8}$$

where the innovation covariance matrix S_k is based on the a priori error covariance matrix $\hat{P}_k$ and the measurement covariance matrix of accelerometer R_a, whose main diagonal elements are from the accelerometer values, and nonmain diagonal elements are all zero conversely. The Kalman gain K_k is the error covariance matrix after gravity measurement is corrected.

3.2.4. Yaw corrections

$$S_k = H_k P_k^{(1)} H_k^T + R_m \tag{9}$$

$$K_k = P_k^{(1)} H_k^{\ T} \left(H_k P_k^{(1)} H_k^{\ T} + R_m \right)^{-1} \tag{10}$$

$$x_k^{(2)} = x_k^{(1)} + K_k \left(m_k - H_k x_k^{(1)} \right) \tag{11}$$

$$P_k^{(2)} = (I_{4\times4} - S_k H_k) P_k^{(1)} \tag{12}$$

where the measurement covariance matrix of magnetometers R_m, which main diagonal elements are from the magnetometer values, non-main diagonal elements are all zero conversely. $P_k^{(2)}$ is the error covariance matrix after yaw measurement is corrected.

3.3. Feature extraction (FE)

AUV motions are defined by the six degrees of freedom, including heave, surge, sway, pitch, roll, and yaw, and they are coupling by the vehicle shape, trends, and current interaction. Therefore, the attitude data of AUV is of high dimension and very complex. The value of FE is to reduce the dimension of the large measurement data and prevent program operation to run out of memories. The characteristics of a data segment are to keep the most meaningful features and remove the redundant data. Therefore, the FE methods have been applied for activity detection from accelerometer data [24, 25]. İn order to extract features easily, the continuous measurement data of sensors are divided into many overlapping segments of which each is 20 seconds long, as illustrated in **Figure 6**. A sliding window technique cuts the sensors' data into 20 seconds (550 samples) in each short-term window with 50% overlapping. As previous studies found that the effectiveness of FE on windows with 50% overlap is an effective window size [24].

In this chapter, we adopted principal component analysis (PCA) as the feature selection procedure to lower the dimension of the original features. Feature extraction is highly subjective in nature, it depends on applications. Here, we introduced the following features that are beneficial to the classification of AUV failure detection, and used these features to discriminate the type of AUV activity: (1) min, max; (2) mean; (3) interquartile range (IQR); (4) root mean square (RMS); (5) standard deviation (STD); (6) root mean square error (RMSE); (7) signal magnitude area (SMA); (8) signal vector magnitude (SVM); and (9) averaged acceleration energy (AEE).

3.4. BPN classifier

Finally, this feature extraction by the PCA is fed into the short-term classifier and long-term classifier sequentially, and then output the AUV conditions. In order to detect the AUV condition accurately, reliably, stably, and robustly, we divided the recognition system into three classifiers which are based on BPN. BPN is considered the workhorse of ANNs and is the multilayer perceptron (MLP) based on a feed-forward algorithm. The hidden layers, between input and output layers, use the error backpropagation (BP) algorithm to compute the nonlinear relationship in supervised learning as shown in **Figure 7**. The main features of BPN are as follows: high learning accuracy, fast response, and ability to process the nonlinear problems.

In this chapter, a three-layer BPN is used for classifying AUV failure conditions. The topology of the BPN classifier is shown in **Figure 7** [26]. The input layer has R neurons, equal to the dimension of the feature vectors $I = \left[I_1^{(1)},\ I_r^{(1)}, \ldots, I_R^{(1)}\right]^T$, where superscript 1 indicates the first layer. The hidden layer has J neurons, and the output layer has P neurons, equal to the number

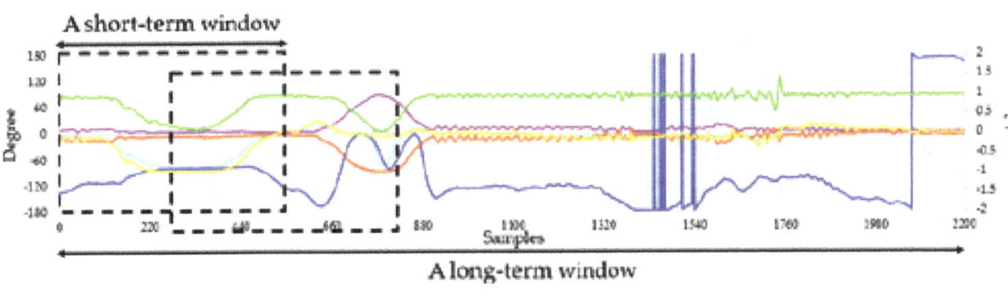

Figure 6. Attitude data with a long-term window, which consists of many overlapping short-term windows.

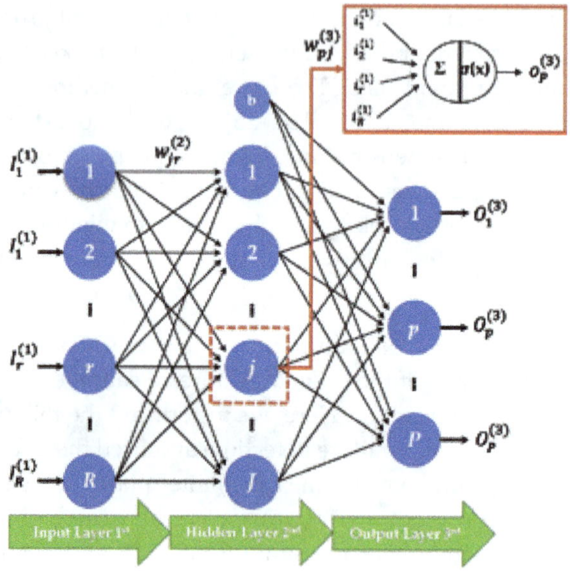

Figure 7. The topology of the BPN classifier.

of AUV conditions $O = \left[O_1^{(3)},\ O_{\mathrm{P}}^{(3)}, \ldots, O_{\mathrm{P}}^{(3)}\right]^T$, where superscript 3 indicates the third layer. $w_{pj}^{(3)}$ and $w_{jr}^{(2)}$ denote the weight from the hidden to the output layer and from the input to the hidden layer, respectively. $b^{(3)}$ and $b^{(2)}$ denote the bias in the third and second layers, respectively. $i^{(\mathrm{n})}$ and $o^{(\mathrm{n})}$ are the input value and the activation of a neuron in the nth layer $\{n \in N:\ n\ \leq 3\}$. We take the interval between the hidden layer (second) and the output layer (third) as an example. The function of the neurons in each step is defined as follows [27]:

In the BPN, the intermediate quantity $net_j^{(3)}$ is the weighted input to the neurons in the third layer, and implements a nonlinear transformation from the output values of the second layer to the output values of the third layer. σ is called the sigmoid function and is in general nonlinear and differentiable. The functions of the pth neuron are given by:

$$o_p^{(3)} = \sigma\left(net_j^{(3)}\right) = \sigma\left(\sum_{j=1}^{J} w_{pj}^{(3)} o_j^{(2)} + b_p^{(3)}\right) \tag{13}$$

The discrepancy $E^{(3)}$ between the desired output $t_p^{(3)}$ and the real output $O_p^{(3)}$ in the third layer can motivate the BP learning algorithm as gradient descent on the sum-squared discrepancy. The factor of 1/2 will simplify the exponent when differentiating later. The function is defined as:

$$\mathrm{E}^{(3)} = \frac{1}{2}\sum_{p}\left(t_p^{(3)} - O_p^{(3)}\right)^2 \tag{14}$$

Then, the weights are adjusted to find the partial derivative E with respect to a weight $w_{\mathrm{pj}}^{(3)}$: $\Delta w_{\mathrm{pj}}^{(3)} \propto -\eta\left(\partial E^{(3)}/\partial w_{\mathrm{pj}}^{(3)}\right)$. However, the discrepancy is not directly the function of weights. Calculating the derivative of the discrepancy is done using the chain rule twice.

$$\Delta w_{pj}^{(3)} = -\eta \frac{\partial E^{(3)}}{\partial o_p^{(3)}} \frac{\partial o_p^{(3)}}{\partial net_p^{(3)}} \frac{\partial net_p^{(3)}}{\partial w_{pj}^{(3)}} \tag{15}$$

Let us discuss each of partial derivatives in turn. In the last term of Eq. (15) is the derivative of the net $net_p^{(3)}$ with respect to a weight $w_{pj}^{(3)}$:

$$\frac{\partial net_p^{(3)}}{\partial w_{pj}^{(3)}} = \frac{\partial}{\partial w_{pj}^{(3)}} \left(\sum_{j=1}^{J} w_{pj}^{(3)} o_j^{(2)} + b_p^{(3)} \right) = o_j^{(2)} \tag{16}$$

Next, the derivative of the activation $o_p^{(3)}$ with respect to the net input $net_p^{(3)}$:

$$\begin{aligned} \frac{\partial o_p^{(3)}}{\partial net_p^{(3)}} &= \frac{\partial}{\partial net_p^{(3)}} \sigma\left(net_p^{(3)}\right) \\ &= \sigma\left(net_p^{(3)}\right)\left(1 - \sigma\left(net_p^{(3)}\right)\right) = o_p^{(3)}\left(1 - o_p^{(3)}\right) \end{aligned} \tag{17}$$

Last, we consider the derivative of E with respect to the activation. Since the neuron is in the output layer, then $O_p^{(3)}$ is equal to $o_p^{(3)}$.

$$\frac{\partial E}{\partial o_{\mathrm{p}}^{(3)}} = \frac{\partial E}{\partial O_{\mathrm{p}}^{(3)}} = \frac{\partial\left(1/2\left(t_{\mathrm{p}}^{(3)} - O_{\mathrm{p}}^{(3)}\right)^2\right)}{\partial O_{\mathrm{p}}^{(3)}} = -\left(t_{\mathrm{p}}^{(3)} - o_{\mathrm{p}}^{(3)}\right) \tag{18}$$

Finally, we substitute these results Eqs. (15)–(17) back into original Eq. (17) to find the weight change $\Delta w_{pj}^{(3)}$ rule,

$$\Delta w_{pj}^{(3)} = \eta \overbrace{\left(t_{\mathrm{p}}^{(3)} - o_{\mathrm{p}}^{(3)}\right) o_{\mathrm{p}}^{(3)} \left(1 - o_{\mathrm{p}}^{(3)}\right)}^{\varepsilon_p} o_{\mathrm{j}}^{(3)} = \eta \varepsilon_p o_{\mathrm{j}}^{(3)} \tag{19}$$

We can find the weight change $\Delta w_{jr}^{(2)}$ in the same processes as above and it is defined as:

$$\Delta w_{jr}^{(2)} = \eta \overbrace{\sum_{j}^{J} \varepsilon_p w_{\mathrm{pj}}^{(3)} o_{\mathrm{j}}^{(2)} \left(1 - o_{\mathrm{j}}^{(2)}\right)}^{\varepsilon_j} o_{\mathrm{r}}^{(1)} = \eta \varepsilon_j o_{\mathrm{r}}^{(1)} \tag{20}$$

The BPN algorithm approach to recognize and intelligently detect failures is based on changes in weight values $\Delta w_{pj}^{(3)}$ $\Delta w_{jr}^{(2)}$ of the AUV feature parameters.

4. Experimental test and result

4.1. SCEKF

To validate the sensor calibration and fusion method proposed in this chapter, this section covers the results obtained from the 10-DOF sensor module with operating system of Windows 8.1, 2.20 GHz CPU, 8 GB memory, and experiments are performed with the Arduino program described in Section 2.1. The system was fixed on an anti-vibration table to minimize interference. The proposed SCEKF algorithm was used to estimate accelerometer and orientation with 27-Hz updating rate.

As the calculation process of sensor algorithm, we used the rotary platform with outputting the quantitative and stable signal to observe the relative signal output. The first process is fixing IMU at the center and along the rotation axis of the platform and fixing it by valves. After installation, setting the required rate or angle of rotation as a reference data by the user interface, we measured the output corresponding to the reference data. The performance of the orientation before/after SCEKF algorithm is presented in **Figure 8(a–c)**, respectively; it produced reasonable output values within the expected ranges. It is seen that each component of Euler orientation (i.e., roll, pitch, and yaw) is within 0·1 degree after SCEKF processing in the static test.

4.2. AUVs' attitude simulation and experiment

Figure 9 shows the whole system of the TrenchRoDer equipped with the proposed AIRS. The whole equipment includes the TrenchRoDer, a camera, an AIRS, a Bluetooth box, and a floating board. We put a camera in the waterproof bags and tied them to a fixed plate in order to observe the situation when the airbag is inflated. The function of the floating board is to protect the Bluetooth box from immersing into the water.

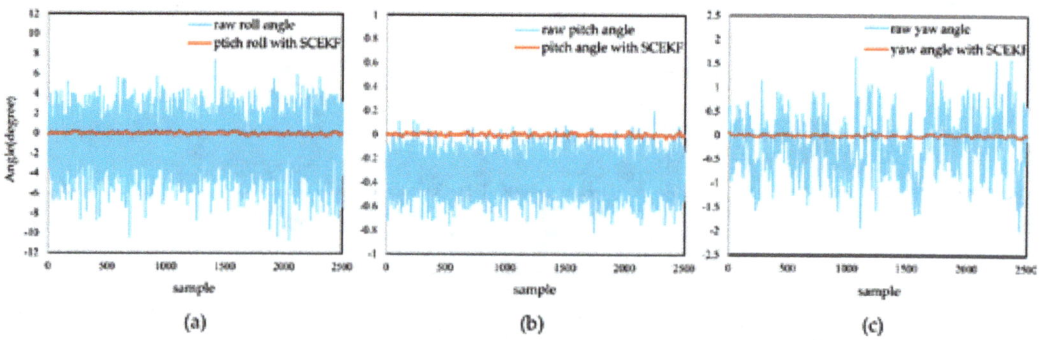

Figure 8. The calibration results from the static condition produced by the SCEKF processing. (a) Roll angle calibration. (b) Pitch angle calibration. (c) Yaw angle calibration.

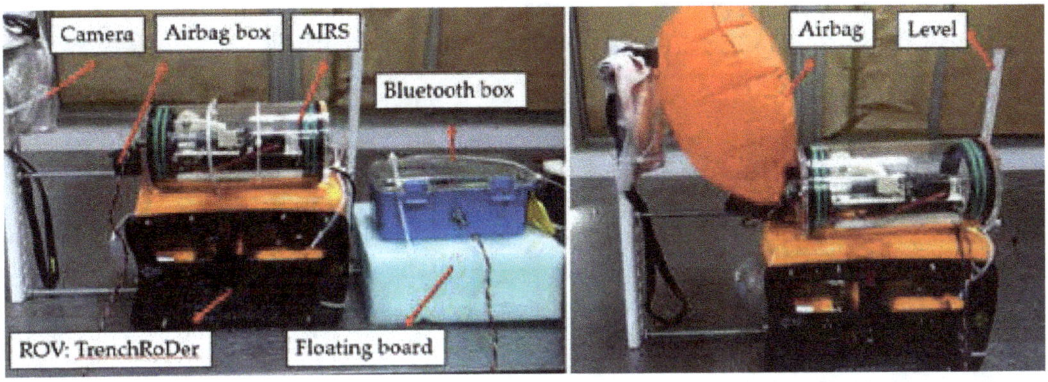

Figure 9. The TrenchRoDer equipped with the proposed AIRS.

In order to construct a robust underwater vehicle fault attitude database, we simulated the underwater vehicle motion model with different situations in the water environment, which can provide the particular case of attitude data, such as if the AUVs suffered a crash or a disabled propeller. Moreover, we built the fault simulator GUI interface for a more convenient operation, in other words, the AIRS starts with the fault simulator GUI interface, after inputting the required AUV's dimensions, hydrodynamic coefficients, buoyancy center, and inertia coefficients. Next, the simulator calculates the different AUV motion data under different cases through the Matlab™ program as shown in **Figure 10**. The above results from simulator will combine with the experimental data as the database for the training and verification of BPN classifier. Last but not least, the real AUV motion signal is set as the testing dataset to ensure the establishment of the AIRS. The modular modeling equation of AUVs is selected from [28]. The modeling method of Prestero's which id without considering the sea conditions of emission can help us to generate the dynamic models of AUVs quickly and conveniently. The dynamics model and kinematics model of AUVs are established by analysis of the force working on AUVs moving underwater, based on the theorem of the momentum of the rotation around the buoyancy center and the theorem of the motion of mass center. The motion in 6-DOF of the AUVs is determined. And then, the attitude data of AUVs at any instant are determined.

Since we simulated the AUV motion in the underwater environment without considering the influence of currents and waves, we did the wave maker experiment in NCKU's ship model towing tank to observe the changes and effects on the AUV motion data for different wave heights. By experimenting with simulation, we can be closer to the state of the real ocean environment. The experiment of the AUVs affected by wave maker in NCKU's ship model towing tank is illustrated in **Figure 11**.

4.3. Classification results

In this chapter, we focused on 10 conditions of AUV, which are listed in **Table 1**.

There are two types of these AUV conditions. One of them is failure situation and the other is functional condition and both include five motion status of the AUV. We carried out 20

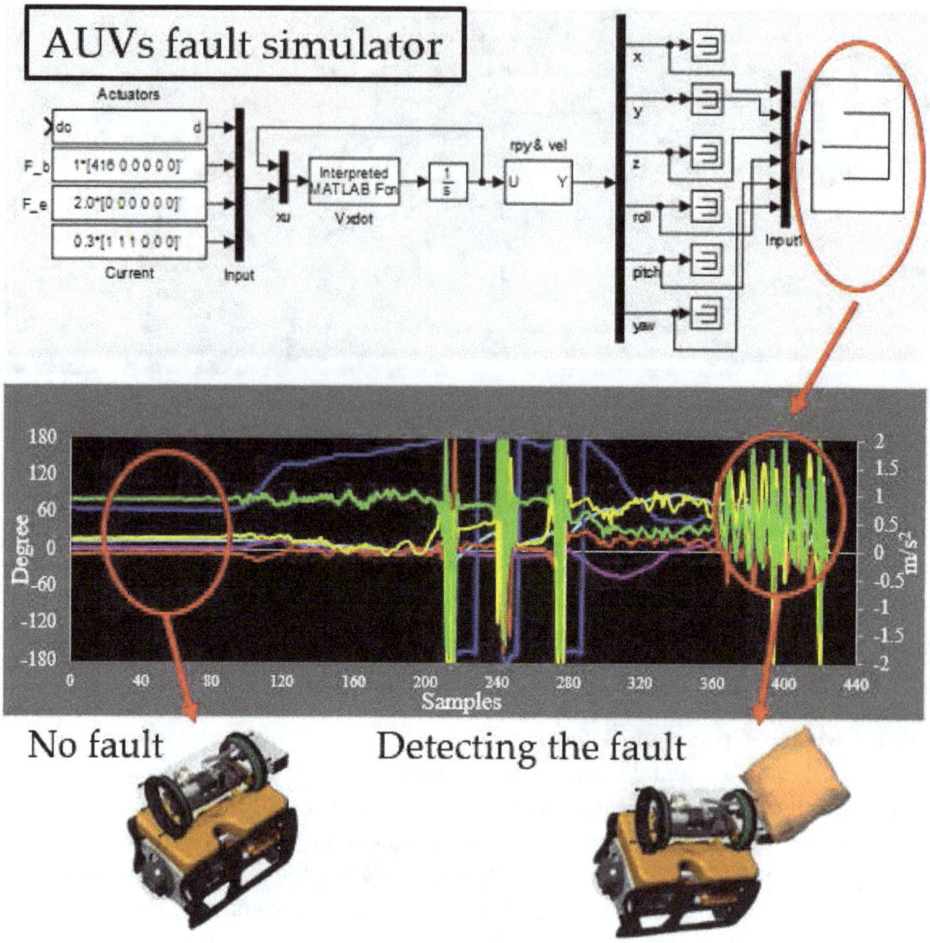

Figure 10. The AUV's fault simulator diagram.

Figure 11. The wave maker of the NCKU's ship model towing tank.

Number	Activity description
1	Functional condition after horizontal sinking
2	Malfunction after horizontal sinking
3	Functional condition after vertical sinking
4	Malfunction after vertical sinking
5	Functional condition after ramped sinking
6	Malfunction after ramped sinking
7	Functional condition after turbulence
8	Malfunction after turbulence
9	Functional condition after collision
10	Malfunction after collision

Table 1. Activities performed in this experiment.

experiments for capturing data and verifying classifier. The data from 17 experiments were adopted in the training program of the recognition scheme; the data obtained from the other experiments were used for testing the recognition performance. Note that, since the sampling frequency is 27.5 Hz, the total number of the short-term and long-term samplings for each activity of each experiment is 550 and 2200, respectively, which means 20 seconds per short-term window and 80 seconds per long-term window. The feature extraction of this chapter was based on 50% overlapping windows using 550 samples of window sizes to avoid information loss at the boundary of a single window. The dimension of a feature vector was 45 (an accelerometer $\times$ 3 axes $\times$ 9 features + a gyroscope $\times$ 3 axes $\times$ 6 features). **Figure 12** illustrates the first 2200 data of accelerations and Euler orientations collected from the first experiment. The selected features of sensor's data enabled effective recognition of the conditions and were suggested for BPN training procedure. A computation program adopted the input features and activated the feature classifier learning procedure with the BP algorithm, and outputted the results to short-term classifier. Then, an AUV condition was distinguished by a long-term classifier, of which the input is from the short-term classifier to raise the accuracy of failure recognition. The number of neurons in each hidden layer is 4, 6, and 7 for the feature classifier, short-term classifier, and long-term classifier, respectively, and the number of epochs is 700 for each neural training. The BPN classifier was trained on the training data set and tested on the test set which are from the experiment values. The classifier was created by neural network toolbox of MATLABTM for practical implementation and for validating the proposed model.

After building up our prediction algorithm, we apply our chosen prediction algorithm on our new test set which is from the real signal of AUV, in order to have an idea about the algorithm's performance on unseen data. The confusion matrix measured in the real AUV test is shown in **Table 2**, which recorded the results from 20 times experiments on each condition of AUV. We have implemented in two different ways under MATLABTM environment. In the first, we conducted in our proposed classifier system with feature extraction and the results indicate that the AUV failure detection on the average 97% of the time, and a successful functional condition accuracy of 93% is achieved. Second, we chose a classifier in ANN learning algorithm without feature extraction for comparing with the classifier that we proposed. The performance indicated

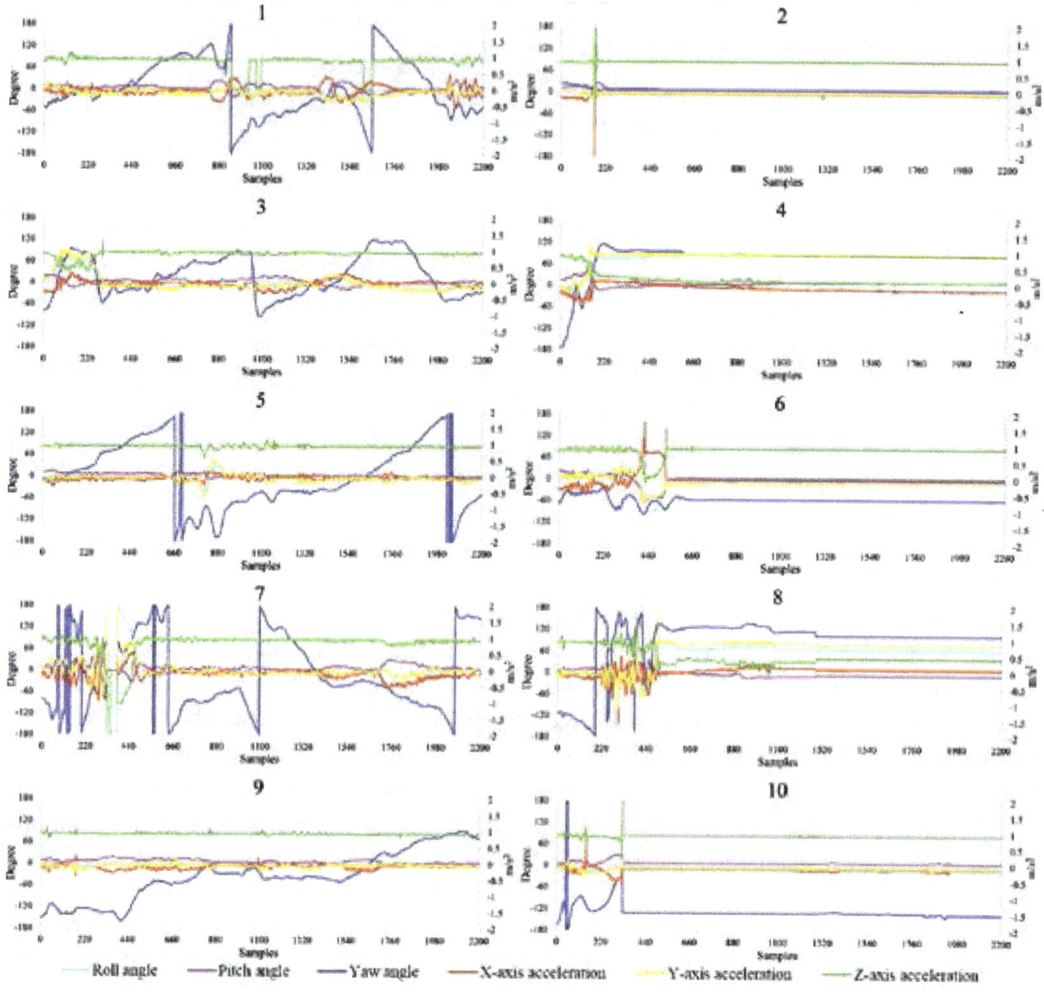

Figure 12. The accelerations and Euler orientation of the first experiment.

20 times of testing experiments													
Classified Type	**1**	**2**	**3**	**4**	**5**	**6**	**7**	**8**	**9**	**10**	**Recognition rate (%)**	**Detection accuracy (%)**	
1	14/13	0/0	1/2	0/0	5/6	0/0	0/0	0/0	0/0	0/1	70/65	Malfunction	97/92
2	1/2	16/14	0/0	0/0	0/0	1/2	0/0	0/1	2/3	2/4	80/70	Functional condition	93/88
3	1/1	0/0	15/13	0/0	0/1	0/0	0/0	0/0	1/1	0/0	75/65		
4	0/0	0/0	0/0	18/16	0/0	2/2	0/1	0/0	0/0	0/0	90/80		
5	4/3	0/0	2/4	0/1	15/13	0/0	0/0	0/0	0/1	0/0	75/65		
6	0/1	1/1	0/0	2/3	0/0	16/15	0/0	3/4	0/0	0/0	80/75		
7	0/0	0/0	0/0	0/0	0/0	0/0	17/15	1/2	0/0	0/0	85/75		

20 times of testing experiments												
Classified Type	**1**	**2**	**3**	**4**	**5**	**6**	**7**	**8**	**9**	**10**	**Recognition rate (%)**	**Detection accuracy (%)**
8	0/0	0/0	0/0	0/0	0/0	1/1	3/4	16/13	0/0	0/0	80/65	
9	0/0	1/2	2/1	0/0	0/2	0/0	0/0	0/0	16/14	1/2	80/70	
10	0/0	2/3	0/0	0/0	0/0	0/0	0/0	0/0	1/1	17/14	85/70	

* Proposed classifier with feature extraction in BP algorithm/ANN classifier without feature extraction.

Table 2. Confusion matrix for all the testing experiments.

that the ANN classifier without feature extraction performs poorer than our proposed classifier. From the confusion matrix, we can know that the malfunction and functional conditions are not easy to be confused. However, the motions within functional condition may be misclassified between each other, because these activities contain similar amplitude peaks and waveforms at the AUV.

5. Conclusions and future works

The main objective of this work was to develop an intelligent AUV rescue system (AIRS) for detecting failure to minimize loss of an autonomous underwater vehicle (AUV). We combined three main ideas to construct the AIRS, including sensors algorithm, classifier conducting, and airbag system. Complex data acquisition was done by the 10-DOF sensor module with sensor calibration and extended Kalman filter (SCEKF), where Euler orientation fused with gravity and magnetic field are state variables, can benefit to get the precise attitude from the AUV. After SCEKF processed, we extracted the features of these signals from the 10-DOF sensor module and selected these features by principal component analysis (PCA) method. The results were incorporated with feature classifier, short-term classifier, and long-term classifier in order to recognize 10 types of AUV conditions. According to the experimental test in Section 3, we have shown that the 20 experimental data sets are categorized into "malfunction" or "functional condition" categories. The outcomes of the proposed classification with features extracted, whose failure detection accuracy is 97%, were more accurate than those of the ANN without features extracted. These results confirmed that the technology of the AIRS was feasible and that the proposed methods were accurate. Furthermore, we designed an inflatable mechanism, which fills CO_2 in the airbags to generate buoyancy for AUV during failure detection.

The attitude estimation and classification applied in the underwater environment are a new field. Considering future work, we will try to extend more condition types of AUV for classifying more complex situations and accomplish the airbag system for setting on the AUV. We can have the ability to construct a variety of different models for the AUV's fault simulation, such as underwater turbulence or underwater creatures' interference. Providing a more comprehensive AIRS, to not have to do the experiment in the real underwater environment, also can improve the performance and convenience of installation to the AIRS.

With this technology, we can install the appropriate number of gas cylinders and airbags in accordance with the different displacements of AUV to avoid the loss of AUV; this can even be used in rescuing vessels to reduce shipwreck in the future, thereby minimizing loss of life and property. This study will have outstanding contributions for the next generation of underwater vehicles. We are looking forward to the application of the SIRS being used widely in the future.

Acknowledgements

The authors hereby extend sincere thanks to the Ministry of Science and Technology (MOST) and Fisheries Agency for their financial support of this research, whose project codes are MOST 106-2221-E-006-121, MOST 107-2218-E-006-031, MOST 107-2218-E-110-004-and 107AS-14.2.7-FA-F1(3) and 107農科-14.2.7-漁-F1(3). We thank to the generous patronage of the MOST and Fisheries Agency that this study has been smoothly performed.

Author details

Sheng-Chih Shen and Yi-Ting Yang*

*Address all correspondence to: grace60518@hotmail.com

Department of Systems and Naval Mechatronic Engineering, National Cheng Kung University, Tainan City, Taiwan ROC

References

[1] Roman C, Mather R. Autonomous underwater vehicles as tools for deep-submergence archaeology. Proceedings of the Institution of Mechanical Engineers Part M-Journal of Engineering for the Maritime Environment. 2010;**224**(M4):327-340

[2] Asakawa K, Kojima J, Kato Y, Matsumoto S, Kato N, Asai T, et al. Design concept and experimental results of the autonomous underwater vehicle AQUA EXPLORER 2 for the inspection of underwater cables. Advanced Robotics. 2002;**16**(1):27-42

[3] Zhou JW, Mason A. Communication buses and protocols for sensor networks. Sensors. 2002;**2**(7):244-257

[4] Hwangbo M, Kim JS, Kanade T. IMU self-calibration using factorization. IEEE Transactions on Robotics. 2013;**29**(2):493-507

[5] Niu X, Zhang Q, Li Y, Cheng Y, Shi C. Using Inertial Sensors of iPhone 4 for Car Navigation; 2012. pp. 555-561

[6] Leardini A, Lullini G, Giannini S, Berti L, Ortolani M, Caravaggi P. Validation of the angular measurements of a new inertial-measurement-unit based rehabilitation system: Comparison with state-of-the-art gait analysis. Journal of Neuroengineering and Rehabilitation. 2014;**11**:7

[7] Yun X, Bachmann ER, McGhee RB, Whalen RH, Roberts RL, Knapp RG, et al. Testing and evaluation of an integrated GPS/INS system for small AUV navigation. IEEE Journal of Oceanic Engineering. 1999;**24**(3):396-404

[8] Bonin-Font F, Massot-Campos M, Lluis Negre-Carrasco P, Oliver-Codina G, Beltran JP. Inertial sensor self-calibration in a visually-aided navigation approach for a micro-AUV. Sensors. 2015;**15**(1):1825-1860

[9] Luinge HJ, Veltink PH. Measuring orientation of human body segments using miniature gyroscopes and accelerometers. Medical & Biological Engineering & Computing. 2005; **43**(2):273-282

[10] Ruiz ARJ, Granja FS, Honorato JCP, Rosas JIG. Accurate pedestrian indoor navigation by tightly coupling foot-mounted IMU and RFID measurements. IEEE Transactions on Instrumentation and Measurement. 2012;**61**(1):178-189

[11] Li W, Wang JL. Effective adaptive Kalman filter for MEMS-IMU/magnetometers integrated attitude and heading reference systems. Journal of Navigation. 2013;**66**(1):99-113

[12] Mirzaei FM, Roumeliotis SI. A Kalman filter-based algorithm for IMU-camera calibration: Observability analysis and performance evaluation. IEEE Transactions on Robotics. 2008; **24**(5):1143-1156

[13] Hopfield JJ. Neural networks and physical systems with emergent collective computational abilities. Proceedings of the National Academy of Sciences of the United States of America-Biological Sciences. 1982;**79**(8):2554-2558

[14] Sayyaadi H, Ura T. Multi Input-Multi Output System Identification of AUV Systems by Neural Network; 1999. pp. 201-208

[15] Zhang GQP. Neural networks for classification: A survey. In: IEEE Transactions on Systems, Man and Cybernetics—Part C: Applications and Reviews. 2000;**30**(4):451-462

[16] Lund AK, Ferguson SA. Driver fatalities in 1985-1993 cars with airbags. Journal of Trauma-Injury Infection and Critical Care. 1995;**38**(4):469-475

[17] Burgess AR, Dischinger PC, Oquinn TD, Schmidhauser CB. Lower-extremity injuries in drivers of airbag-equipped automobiles: Clinical and crash reconstruction correlations. Journal of Trauma-Injury Infection and Critical Care. 1995;**38**(4):509-516

[18] Wang L. Air bags for motorcycles. Chemical & Engineering News. 2005;**83**(39):80-80

[19] Dragcevic Z, Takeuchi K, Vecaj D, Hursa A. Motorcycle jacket with integrated air bag. Tekstil. 2009;**58**(7):346-351

[20] Guangyi S, Cheung-Shing C, Guanglie Z, Li WJ, Leong PHW, Leung KS. Towards a Mobile Airbag System Using MEMS Sensors and Embedded Intelligence; 2007. pp. 634-639

[21] Toshiyo T, Takumi Y, Masaki S. A Study to Demonstrate the Use of an Air Bag Device to Prevent Fall-Related Injuries; 2008. pp. 1-3

[22] Tamura T, Yoshimura T, Sekine M, Uchida M, Tanaka O. A wearable airbag to prevent fall injuries. IEEE Transactions on Information Technology in Biomedicine. 2009;**13**(6):910-914

[23] Ishizaka S, Moromugi S, Kobayashi M, Kajihara H, Koga K, Sugahara H, et al. A remote-controlled airbag device can improve upper airway collapsibility by producing head elevation with jaw closure in normal subjects under propofol anesthesia. IEEE Journal of Translational Engineering in Health and Medicine. 2014;**2**:1-9

[24] Ravi N, Dandekar N, Mysore P, Littman ML. Activity recognition from accelerometer data. In: AAAI. 2005;**5**:1541-1546

[25] Saeedi S, El-Sheimy N. Activity recognition using fusion of low-cost sensors on a smartphone for mobile navigation application. Micromachines. 2015;**6**(8):1100-1134

[26] Yang JY, Wang JS, Chen YP. Using acceleration measurements for activity recognition: An effective learning algorithm for constructing neural classifiers. In: Pattern Recognition Letters; 2008. pp. 2213-2220

[27] Rumelhart DE, Hinton GE, Williams RJ. Learning internal representations by error propagation. In: Parallel distributed processing: Explorations in the microstructure of cognition, California University San Diego La Jolla Inst for Cognitive Science. 1986;**1**:318-362

[28] Prestero TTJ. Verification of a Six-degree of Freedom Simulation Model for the REMUS Autonomous Underwater Vehicle. In: PhD Thesis. Doctoral Dissertation, Massachusetts Institute of Technology; 2001

6

Alternative Position Estimation Systems for Micro Air Vehicles

Mathias Fassini Mantelli

Abstract

Micro air vehicles (MAVs) is a technology that is becoming more and more important and popular nowadays. It is used as a tool to deal with different tasks that were not possible in the past. For most MAV models, the GPS sensor is the only way of estimating its pose in the environment. However, besides the fact of not having a secondary position estimation system besides the GPS, this is also risky because the GPS may fail like any other sensor. To overcome this weakness and make the MAVs more robust to autonomous tasks, the research community proposed many different localization systems for different constraints. In this chapter, the most popular, recent, and important MAV localization systems are reviewed, as well as the promising future works in this field.

Keywords: MAVs, GPS, localization system, maps, images

1. Introduction

One of the first uses of micro air vehicles (MAVs) was during the World War I [1]. Since then, MAVs have been considered as a promising technology, and nowadays they are being used in several different tasks, such as agriculture [2], patrolling [3], mapping [4], and delivering [5]. Compared to conventional human-crewed aerial vehicles, MAVs are a low-cost and entirely suitable alternative for repetitive or high-precision demanding tasks. Besides, they are also recommended for low-altitude flights and for those that demand a high range of maneuvers.

The estimation of the MAVs' position, i.e., its localization in the world, is the main common requirement between all the before mentioned tasks, even if they would be addressed by other types of mobile robots rather than MAVs. For such complex tasks, localization and navigation are fundamental capabilities that allow MAVs to accomplish their mission [6]. The localization for MAVs is usually solved by the global positioning system (GPS) [7], in which an embedded GPS sensor communicates with different satellites that are orbiting the Earth to estimate the position. Other MAV models rely on different ways of estimating their position, such as inertial sensors or visual odometry algorithms. However, most of these models do not have a second position estimation system in addition to the primary system, i.e., the GPS [7]. Therefore, the MAVs that depend exclusively on GPS to estimate their position are more likely to fail on their tasks or missions; once like any other sensor, the GPS might fail, and they do not have a redundant position estimation system.

Even though it is widely used in different situations and for distinct goals, the GPS sensor is vulnerable to some problems [8, 9]. The amount of satellites that are available to establish a communication with the GPS influences the position estimation certainty, as well as the signal quality between them. The signal might be affected by the weather, such as cloudy and rainy days, and by obstacles, like high buildings or hills. Hence, the higher is the number of connected satellites and the stronger is the signal, the lower is the GPS position estimation error. In addition to this GPS weakness, there is another problem that might disturb the GPS position estimation, the so-called Jamer guns. While the GPS sensor is reading the satellite signal to estimate the position, these guns jam the signal, and hence, the estimation becomes unreliable [10, 11].

The mobile robotics research community has investigated the MAV position estimation problem, and valuable works have been proposed. In general, it is addressed by them as the localization problem from the mobile robotics field, in which the goal is to estimate the pose of a robot, based on readings of its sensors, in an a priori known map [12]. The works proposed by the community covers a considerable variety of approaches, in which the main differences are the kind of data used to represent the environment and the technique used to estimate the pose. Despite this diversity, one characteristic that most of them share is the use of visual data from cameras to estimate the localization. This choice is made due to the advantages of cameras to deal with this problem in comparison to the other sensors, such as the low weight for MAVs, the distinct information from one image (color, depth, intensity, etc.), and the long-distance range for the readings.

This chapter covers the most important proposed works that aimed to deal with visual MAV localization problem. As aforementioned, there are two main topics that are worthy to be covered when presenting this kind of works, which are the data used as a map and how the estimation is calculated. Therefore, this chapter first presents a discussion about different maps used so far, followed by the review of the localization itself. In addition to detailing and comparing them, it also presents what the next trends or future work for this problem are.

2. Localization problem

Mobile robots aiming to perform tasks without human interference, i.e., autonomously, must know their pose within the environment. The same necessity applies for MAVs that have only GPS sensor as position estimation. Estimating the robot's pose would be a simple task, but only if all the sensors of the robot were perfect and the environment fully static. Given that this scenario is not realistic, in which the sensor readings are not precise and many agents are moving through the environment, the difficulty level of the localization problem increases, and hence, it is necessary to estimate the robot's pose.

Despite its difficulty level, localization is a fundamental problem in the mobile robotics field [13]. It is defined as the robot's pose estimation relative to a previously known environmental map [12]. Even though its definition is simple, this problem has two main variations: local and global estimations. In the former, the initial robot pose in the map is known, and the local localization approach only tracks the robot as it moves through the environment. The error of the first pose estimation is low, and the goal is to keep it low using the sensor readings and the motion information from the robot. In contrast to the former variation, the second one is significantly harder. In this case, the initial robot pose is unknown, and hence, the error of the pose estimation is originally high. Instead of considering just a small part of the map at the beginning, in the global localization, the whole map must be considered for the estimation since the initial pose is unknown [14]. **Figure 1** illustrates the differences between local and global localization. The global localization is illustrated in **Figure 1(a)**, in which the error estimation is high at the begging, and the goal is to reduce it as the robot moves through the environment. The opposite happens in **Figure 1(b)**, which depicts the local localization. The error estimation begins considerably low, and even though it increases through time, as well as the global localization, the goal is to reduce it.

The most popular approaches that deal with localization in the mobile robotics field are grouped either as probabilistic or as deterministic. In the first group, there are two main approaches that are worth it to be mentioned, Kalman filter [15] and particle filter (Monte Carlo) [16]. Even though both implement the Bayes filter, each one has its specific advantages, and therefore, they are suitable for different situations and constraints. On the other hand, the most popular approach for the second group is based on interval analysis, and the estimation is defined through boxes that must be minimized [17]. As the goal of this chapter is not to go deep into these approaches, the reader is invited to look at the references for more details about them.

Independent of the approach used to deal with the localization problem, all of them have the same characteristics: as input, they require an environment representation, a sensor to read the environment, and odometry data; and as output, the robot's position in the environment representation that was estimated, as shown by **Figure 2**. Hence, localization systems try to find the best pose in the map that fits both the sensor reading and the odometry information. The best the system is, the more accurate is the pose estimation. **Figure 2** presents an example of using Lidar and 2D map, but it is important to highlight that the same idea also applies to other sensors or types of maps.

Even though the setup of the mobile robot localization problem seems quite simple, with input and output well defined, the difficulty level is considerably high. The robot's pose is not sensed directly, it must be estimated, and that is where the problem lies. Usually, the robot's sensors, both to read the environment and to measure the odometry, are noisy, and hence, the data that they provide does not correctly represent the real world. In addition to that, some types of robots have restrictions about which kind of sensor they support, and they can not have the

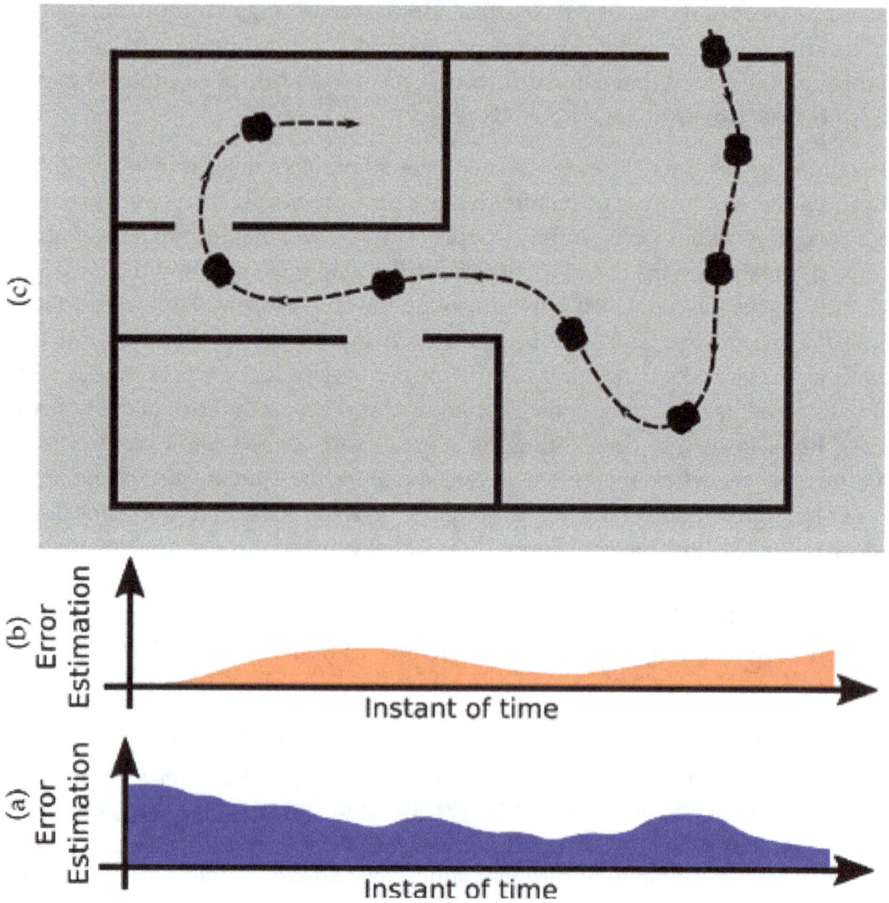

Figure 1. Comparison between the error estimation of (a) global and (b) local localization methods. Both localizations were made considering the robot movement in a (c) 2D map.

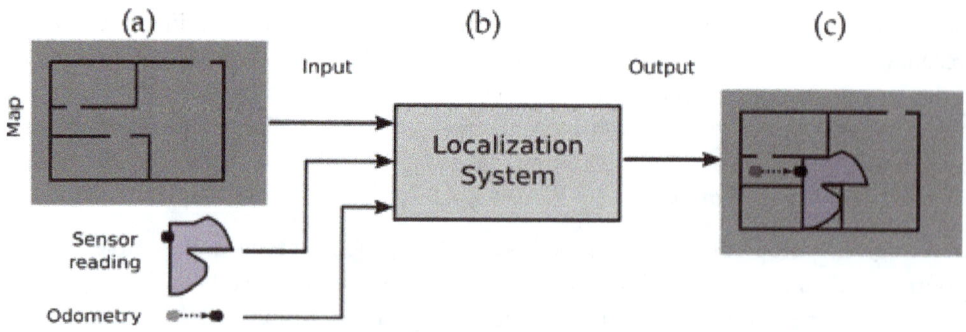

Figure 2. The concepts of a localization system, in which a 2D map, sensor reading, and odometry information are used as input in (a), processed by the system in (b), and the estimated robot's pose is the output in (c).

best sensor for their tasks. In MAVs, for example, a camera is the most popular sensor used for this purpose, since they are smaller, lighter, and cheaper than most range-finder lasers, for instance. However, using images to estimate the odometry information is not ideal, although there are algorithms that compute this estimation.

The localization approaches overcome the noise data problem modeling the error of the sensors. Besides, since just one reading is insufficient to the pose estimation, these approaches also have to integrate the data over time to reduce the error estimation. In environments that have different regions that look alike, such as a building with many corridors and doors, it is quite impossible to estimate the robot's pose considering just one reading. For example, imagine that at some point, the robot is observing a door after having observed a wall and a frame. Then, instead of searches for all the spots in the map that contain a door, the localization system searches for spots that also matches with the wall and the frame. In this way, the past observations are also considered when estimating the robot's pose.

Despite the generic localization problem explanation presented so far, the research community has explored the UAV localization problem throughout the years, and many different approaches have been proposed. The most significant difference between them relies on the map representation and also on the method that they use to compare the sensor readings and the small parts of the map. The next section covers the most popular proposed approaches aiming to deal with this problem and how they differ from each other.

2.1. Alternatives for UAV localization systems

In this section, we review the most important works proposed to deal with UAV localization problem. First, this section presents an analysis of the environment representation that these works used as a map, their advantages and disadvantages. Second, what these works use to compare the sensor readings with different parts of the map. Then, **Table 1** in this section introduces other qualitative comparisons between these works, such as the single or multi MAV pose estimation and indoor or outdoor localization.

2.1.1. Environment representation

The first environment representation presented here is the 2D satellite image map. In general, it is downloaded beforehand from any imagery map source, either entirely or divided into many small images to be stitched later, and then used by the localization approaches to the pose estimation [18, 21, 26, 31]. Even though the research community does not that much explore it, it is also possible to fly the MAV over the region of interest and build the 2D image from the environment, to then use it as a map for the localization estimation. Also, another common choice between the approaches that adopt this kind of map is to point the MAV camera downwards. Then, the MAVs images are compared to different patches from the satellite image map by different comparison methods. The advantages of using the 2D satellite image as a map are the free access to this kind of data through Google Maps or any geographical imagery system (GIS), the excellent representation of the environment by colorful images, and the world coverage. Usually, a GIS also provides the geographic coordinates of satellite images, and then, it is possible to infer the latitude and longitude of each pixel of that image. Therefore, at the same time that a localization system estimates in which pixel from the satellite image map the MAV's pose is, it also estimates the pose in relation to the world, due to the latitude and longitude information in the pixel. In contrast, the disadvantages of such kind of map are the limited point of view (2D) and that some places of the world are not often visited by satellites, and hence, the images are not updated. The comparison methods from the works mentioned above are proposed aiming to be robust against such differences between the outdated 2D satellite images and the MAV image [18, 21, 26, 31].

The disadvantage of a limited point of view from the 2D satellite images motivated the researches to investigate the benefits of 3D maps [19, 20, 23, 30]. The authors argue that by using 3D maps, it is possible to take advantage of the environment structures to estimate the localization, besides the color of the map. Usually, the localization estimation is made based on the 3D structure alignment or even the point cloud matching. For the 3D map case, the MAV camera can be set in different angles, exploring different sights. As well as the 2D satellite images, this kind of 3D representation can be either built right before the localization estimation, as done by the works [20, 23], or downloaded from a GIS, as the case of [19]. Despite these advantages in comparison to the 2D maps, 3D maps generally allocate more computational resources than the 2D one, both to be stored and manipulated, and is not as easy to be found as the 2D maps, what limits the places that it is possible to estimate the MAV's pose.

It is important to highlight that even though flying the MAV before the localization estimation to build the map provides a certainly updated map, for both 2D and 3D ones, this option presents a trade-off. First, a human must pilot the MAV to gather 2D or 3D data from the environment, to then submit it to a mapping approach. Second, it demands more time to start the localization algorithm, since flying the MAV over an area takes more time than downloading a map from a GIS.

In contrast to these two types of maps that represent the whole environment, other map options are more straightforward in terms of details and what is represented. Instead of having a map illustrating all the obstacles, free spaces, and etc., these simple maps only show the position of a few markers. In this case, the idea is to measure the distance between the MAV and all the markers within the map and then estimate the MAV's pose. The type of the markers also varies considerably, such as the case of WLAN access points [28], which are fixed in some spots of the environment and whose received signal strength is measured as part of the localization estimation, and ultraviolet LED markers [24], which emit light in frequencies that are less common in nature than the visible light or infrared radiation. Then it increases the precision of the distance measurement. In this work, the LED markers are not fixed, they are embedded in every MAVs, and they have a mutual relative localization [24]. In more details, they estimate a MAV's pose to another MAV, instead of the global coordinate system. Another marker that it is worth to be mention is the use of tether [27]. The tether reel is fixed in a specific position, and the tether is attached to the MAV. For this case, the MAV is localized to the tether reel by using mechanics model. In general, the use of markers map is adopted for indoor localization, since the sensors that measure the markers have a more limited range than cameras for the 2D or 3D maps presented earlier. The work that relies on ultraviolet LED markers is one exception for this indoor limitation, but this occurs because the MAV's pose is the estimation concerning other MAVs, not to the environment.

Besides the maps presented in this section, other types were tested in the MAV localization problem by the research community. However, they are really specific for a kind of sensor or configuration, and our goal here is to cover the most popular and recent ones. About the types of maps presented here, each one has its advantages and disadvantages, as well as its specific constraints that fit better in some situations. For instance, the 2D satellite image is available online and free but is not suitable for indoor localization. On the other hand, markers map is the option that is most used for indoor localization, but usually, it requires many markers spread through the environment, and it has a short range to be detected. Given that the map of a MAV localization system is essential for the estimation, the type of the MAV, the environment, and the embedded sensor must be taken into account to choose the type of map that fits the constraints better.

2.1.2. Localization estimation methods

In addition to the environment representation, i.e., the map, the methods that estimate the MAV's pose also play an important role in localization systems. Usually, they receive as input the map of the environment and the sensor reading, and then the goal is to find the part of the map that best matches with the sensor reading. In **Figure 2**, the localization estimation method is within the localization system, **Figure 2(b)**, and together with the motion model, it is possible to properly estimate the MAV's pose, **Figure 2(c)**.

Different from the previous section, which introduced the maps used by the MAV localization systems in big groups, such as the 2D satellite images or the marker maps, the localization estimation methods do not share the same similarity between themselves. Then, here they are discussed individually, and as well as in the previous section, their advantages and disadvantages are highlighted.

It is natural that the works that rely on 2D satellite image as a map have an estimation method that is based on image comparison, since their sensor readings are images. The general idea is to compare every MAV image with different patches from different poses within the map, and the most similar one probably represents the MAV's pose in the map. To extract the so-called patches from the map, some global localization works use the Monte Carlo algorithm to extract patches from the whole map [18, 31], whereas the local localization ones, given that the MAV's initial pose is known, only extract a patch from the initial pose and keep extracting them as it moves through the environment [21, 26]. When comparing the MAV images against the patches, each Monte Carlo-based work proposed a novel measurement model, in which one has a new image descriptor called abBRIEF to robustly compute an image signature to the comparison [31], and the other used SURF descriptors [32] and machine learning to compare MAV images and the patches [18]. Both approaches can compute the similarity between all pairs of MAV image and patch and find the most similar pair. On the other hand, the other local localization approaches have a reduced search space, since they know the initial patch from the satellite image. In this case, the image comparison is mainly made by two methods:

either by template matching [22], in which a pixel-by-pixel comparison is performed between two images, or by feature matching [26], that involves the detection, extraction and matching of the features from the two images. In general, given that all these works rely on an image, they use either features or image descriptors to represent the images and then perform the comparison. In their proposal, they aim to overcome the problem of illumination and color changes that occur when dealing with images and outdoor environment, as the case of MAV localization systems.

The alignment technique is also used by other approaches to estimate the MAV's pose, even the ones that rely on 3D maps. In [23], the alignment is made considering the 2D keypoints from the MAV image and the 3D landmarks of the 3D map. To do so, the authors cluster the landmarks into visual words to speed up the matching and alignment with a nearest neighbour search. This 2D to 3D alignment, or transformation, is also applied in another work [19]. Given that the map in this work is a 3D representation of the environment, but the MAV image is a simple 2D RGB image, they have to transform the MAV image into a 3D data, to then align the lines and edges detected in both. As these both works perform local localization, they have an initial reduced search space, which helps them to have a good alignment at the beginning.

Besides the estimation methods presented so far, other ones are even more specific. In [25], for instance, a robust and quick response landing pattern is designed to be visually detected through images and then assist the MAV to its landing. In such a case, the pattern is the map, and the computer vision method proposed by the authors of this work can detect the scale of the map and then estimate the MAV localization. In [24] a markers detection-based approach is also proposed to estimate the MAV's pose. However, in contrast to [25], in [24] the markers are ultraviolet LED that are embedded in the MAVs. Hence, the estimation, in this case, is a mutual one, i.e. one MAV estimates its pose in relation to another one and vice versa. First, their algorithm detects the size of the markers in the image, and then it estimates the internal distance between a pair of markers. Therefore, they can calculate the distance between two MAVs and their pose. In addition to that, in [27] tether-based feedback and inertia sensing are used to estimate the MAV's pose. In more details, the length, azimuth, and elevation angle of the tether are the input for a mechanics model that calculates the absolutely straight tether between the origin and the MAV. The work [28] also relies on a non-popular sensor to estimate the MAV's pose, and the goal is to detect access points (AP) and measure the received signal strength. Then, they can estimate the MAV's pose relative to the APs that have their positions well defined in the map. A similar approach is proposed in [30], in which the MAV's pose is estimated in an urban environment by the transmission of beacons. They are located in different buildings, and they provide a local frame of reference, supporting the MAVs for their location estimation by providing the details of the area and height of the buildings. It is also possible to say that sonar is another type of sensor not easily found embedded on MAVs, and this is the sensor used in [29]. To estimate the MAV's pose, the authors proposed a multi-ray model based on the four sonars sensors embedded in a MAV. This model approximates a beam pattern accurately, and it does not require high computational power.

In general, the localization estimation methods are responsible for comparing the sensor reading and a sample of the map. In another way, it is also known as a transformation from the local coordinate system, i.e., the robot sensor reading, to a global one, i.e., the map. Given the works presented in this section, we can notice that sometimes the sensor reading and the map are different data, such as 2D images from a regular RGB camera and a 3D map. Also, in some cases, these methods have to estimate the MAV's pose based on an outdated map. Because of that, they have to be robust against differences between the real and the mapped environment, and even though they might not represent the same area, the pose should be estimated.

2.1.3. *Summarization*

Table 1 compiles the information presented in Section 2.2. This review shows how the type of map and the sensor changes from system to system. Even though some approaches seems very similar, such as [18, 31], they are still different. Hence, they have their own advantages and disadvantages.

2.2. The future of UAV localization systems

Despite the great effort of the research community to deal with the MAV localization problem, there is no solution that works for all the possible environments and constraints. This problem varies considerably, such as the environment, which can be either indoor or outdoor, the knowledge about the initial position, whether it is known or not, and the amount of MAVs that are being localized, which can be single or multiple MAVs. Despite these difficulties, there is also the map issue, which is caused either by a low updating frequency, such as a satellite that takes some time to revisit a specific area, or by a quick environmental change, like a snowy day that makes the whole environment become white.

To deal with the problem of the type of environment, the approaches that seem more likely to work are the one that recognizes the objects within the environment and the other that deals

Paper	Type of localization	Type of map	Indoor or outdoor	Sensor	Multiple or single MAV
[28]	Local	Access points	Indoor	RSS WLAN	Single
[29]	Local	2D map	Indoor	Sonars	Single
[21]	Local	2D Satellite image	Outdoor	Camera	Single
[26]	Local	2D Satellite image	Outdoor	Camera	Single
[23]	Local	Point cloud	Outdoor	Camera	Single
[19]	Local	3D map	Outdoor	Camera	Single
[25]	Local	Landing pattern	Outdoor	Camera	Single
[27]	Local	Tether reel	Both	Tether	Single
[24]	Mutual	LED markers	Both	Camera	Multiple
[30]	Global	3D map	Outdoor	LFR reader	Multiple
[31]	Global	2D Satellite image	Outdoor	Camera	Single
[18]	Global	2D Satellite image	Outdoor	Camera	Single

Table 1. Comparison between all the discussed works in terms of different information.

with 3D structures. Hence, independently of being indoor or outdoor, it would be possible to recognize objects or detect the environment's shape in both scenarios, to then continue the pose estimation calculation. On the other hand, the problem of the outdated map could be overcome by using deep learning, which provides robust solutions for different seasons or illumination changes in images [33]. Another solution that is possible through the use of deep learning is to teach a net to differentiate roads, buildings, and forest, to then segment both the map and the MAV sensor readings. Hence, instead of, for instance, matching the color of different pixels from MAV images and patches from the 2D satellite image map, the matching would be done considering the classes of the environment, avoiding the problems caused by color or illumination changes.

Due to the fact that MAV is a certainly popular technology and that it is being used in many different tasks, another promising matter that should be investigated is the localization system for multiple MAVs. As there will be even more MAVs flying and cooperating in the future, it is essential to have localization approaches that take advantage of the high amount of MAVs available in the air and, therefore, improve the pose estimation.

Author details

Mathias Fassini Mantelli

Address all correspondence to: mathiasfassini@gmail.com

Federal University of Rio Grande do Sul, Porto Alegre, Brazil

References

[1] Shaw IGR. The rise of the predator empire: Tracing the history of U.S. Drones. Understanding Empire. 2014. Available from: https://understandingempire.word press.com/2-0-a-brief-history-of-us-drones/

[2] Tokekar P, Vander Hook J, Mulla D, Isler V. Sensor planning for a symbiotic UAV and UGV system for precision agriculture. IEEE Transactions on Robotics. 2016;**32**(6):1498-1511

[3] Girard AR, Howell AS, Hedrick JK. Border patrol and surveillance missions using multiple unmanned air vehicles. In: 43rd IEEE Conference on Decision and Control (CDC), Vol. 1; 2004 December 14; 2004. pp. 620-625

[4] Nex F, Remondino F. UAV for 3D mapping applications: A review. Applied geomatics. 2014;**6**(1):1-5

[5] Haque MR, Muhammad M, Swarnaker D, Arifuzzaman M. Autonomous quadcopter for product home delivery. In: 2014 International Conference on Electrical Engineering and Information & Communication Technology; 2014 April 10; 2014. pp. 1-5

[6] Senlet T, Elgammal A. Satellite image based precise robot localization on sidewalks. In: 2012 IEEE International Conference on Robotics and Automation; 2012 May 14; 2012. pp. 2647-2653

[7] Zheng M, Wu C, Chen D, Meng Z. Rotation and affine-invariant SIFT descriptor for matching UAV images with satellite images. In: Proceedings of 2014 IEEE Chinese Guidance, Navigation and Control Conference; 2014 August 8; 2014. pp. 2624-2628

[8] Carroll J. Vulnerability assessment of the U.S. transportation infrastructure that relies on the global positioning system. Journal of Navigation. 2003;**56**(2):185-193

[9] Caballero F, Merino L, Ferruz J, Ollero A. Improving vision-based planar motion estimation for unmanned aerial vehicles through online mosaicing. In: IEEE International Conference on Robotics and Automation. Orlando, FL: ICRA; 2006. pp. 2860-2865

[10] Conte G, Doherty P. An integrated UAV navigation system based on aerial image matching. In: 2008 IEEE Aerospace Conference; 2008. pp. 1-10

[11] Viswanathan A, Pires BR, Huber D. Vision-based robot localization across seasons and in remote locations. In: 2016 IEEE International Conference on Robotics and Automation (ICRA); 2016. pp. 4815-4821

[12] Thrun S, Burgard W, Fox D. Probabilistic Robotics. MIT Press; 2005

[13] Makarenko AA, Williams SB, Bourgault F, Durrant-Whyte HF. An experiment in integrated exploration. In: IEEE/RSJ International Conference on Intelligent Robots and Systems, Vol. 1. 2002. pp. 534-539

[14] Se S, Lowe D, Little J. Local and global localization for mobile robots using visual landmarks. In: International Conference on Intelligent Robots and Systems. Expanding the Societal Role of Robotics in the Next Millennium, Vol. 1. 2001. pp. 414-420

[15] Leonard JJ, Durrant-Whyte HF. Mobile robot localization by tracking geometric beacons. IEEE Transactions on Robotics and Automation. 1991;**7**(3):376-382

[16] Dellaert F, Fox D, Burgard W, Thrun S. Monte Carlo localization for mobile robots. ICRA. 1999;**2**:1322-1328

[17] Kieffer M, Jaulin L, Walter E, Meizel D. Robust autonomous robot localization using interval analysis. Reliable Computing. 2000;**6**(3):337-362

[18] Masselli A, Hanten R, Zell A. Localization of unmanned aerial vehicles using terrain classification from aerial images. In: Intelligent Autonomous Systems 13. Cham: Springer; 2016. pp. 831-842

[19] Qiu K, Liu T, Shen S. Model-based global localization for aerial robots using edge alignment. IEEE Robotics and Automation Letters. 2017;**2**(3):1256-1263

[20] Lim H, Sinha SN, Cohen MF, Uyttendaele M. Real-time image-based 6-dof localization in large-scale environments. In: 2012 IEEE Conference on Computer Vision and Pattern Recognition. IEEE; 2012. pp. 1043-1050

[21] Yol A, Delabarre B, Dame A, Dartois J, Marchand E. Vision-based absolute localization for unmanned aerial vehicles. In: 2014 IEEE/RSJ International Conference on Intelligent Robots and Systems. Chicago, IL: IEEE; 2014. pp. 3429-3434. DOI: 10.1109/IROS.2014.6943040

[22] Dame A, Marchand E. Second-order optimization of mutual information for real-time image registration. IEEE Transactions on Image Processing. 2012;**21**(9):4190-4203

[23] Surber J, Teixeira L, Chli M. Robust visual-inertial localization with weak GPS priors for repetitive UAV flights. In: 2017 IEEE International Conference on Robotics and Automation (ICRA). Singapore: IEEE; 2017. pp. 6300-6306. DOI: 10.1109/ICRA.2017.7989745

[24] Walter V, Saska M, Franchi A. Fast mutual relative localization of UAVS using ultraviolet LED markers. In: 2018 International Conference on Unmanned Aircraft Systems (ICUAS); Dallas, TX: 2018. pp. 1217-1226. DOI: 10.1109/ICUAS.2018.8453331

[25] Yuan H, Xiao C, Xiu S, Zhan W, Ye Z, Zhang F, et al. A hierarchical vision-based UAV localization for an open landing. Electronics. 2018;**7**(5):68

[26] Goforth H, Lucey S. GPS-Denied UAV Localization using Pre-existing Satellite Imagery. In: 2019 International Conference on Robotics and Automation (ICRA). Montreal, QC, Canada: IEEE; 2019. pp. 2974-2980. DOI: 10.1109/ICRA.2019.8793558

[27] Xiao X, Fan Y, Dufek J, Murphy R. Indoor uav localization using a tether. In: 2018 IEEE International Symposium on Safety, Security, and Rescue Robotics (SSRR). IEEE; 2018. pp. 1-6

[28] Zhou M, Yuan H, Wang Y, Tan W, Tian Z. Indoor UAV localization using manifold alignment with mobile AP detection. In: ICC 2019—2019 IEEE International Conference on Communications (ICC). Shanghai, China: IEEE; 2019. pp. 1-6. DOI: 10.1109/ICC.2019.8761348

[29] Yang L, Feng X, Zhang J, Shu X. Multi-ray modeling of ultrasonic sensors and application for micro-UAV localization in indoor environments. Sensors. 2019;**19**:1770

[30] Sharma V, DNK J, You I, Kumar R, Li J. Secure and efficient context-aware localization of drones in urban scenarios. IEEE Communications Magazine. 2018;**56**(4):120-128

[31] Mantelli M, Pittol D, Neuland R, Ribacki A, Maffei R, Jorge V, et al. A novel measurement model based on abBRIEF for global localization of a UAV over satellite images. Robotics and Autonomous Systems. 2019;**112**:304-319

[32] Bay H, Tuytelaars T, Van Gool L. Surf: Speeded up robust features. In: European Conference on Computer Vision. Berlin, Heidelberg: Springer; 2006. pp. 404-417

[33] Liu M-Y, Breuel T, Kautz J. Unsupervised image-to-image translation networks. In: Advances in Neural Information Processing Systems. Curran Associates, Inc.; 2017. pp. 700-708

7

Review of Autonomous Underwater Vehicles

Kondeti Lakshmi Vasudev

Abstract

The exploration of ocean space requires underwater vehicles (UV) such as submarines, autonomous underwater vehicles (AUV), manned underwater vehicles, remotely operated vehicles (ROV) and ship towed instrumentation packages. Of these, AUVs dominate the exploration of deep oceans. The list of applications where UVs can be employed include long-term deployments where they would serve as platforms for spatiotemporal samplings of physical characteristics (e.g., temperature, depth, conductivity, current) of the water column; use of multiple vehicles for mapping out an evolving phenomenon such as hydrothermal vents, tsunamis, etc., rapidly; transiting long distances to a site for making observations as part of a response team; search and mapping of seabed minerals; underwater warfare using submarines; and mine hunting, pipe laying, and inspection and maintenance of offshore structures.

Keywords: computational fluid dynamics, autonomous underwater vehicles, hydrodynamic drag, wake fraction

1. Introduction

UVs can be classified in a variety of ways. Submarines are manned underwater vehicles. Their operating speed varies from 8 to 20 m/s and depth of operation varies from 200 to 600 m. The length of submarines in existence varies from 57.3 m (Dolphin 1 class) to 175 m (Typhoon class). Submarines are used in underwater warfare, covert operations, and coastal defense. AUVs are underwater robots with operational speeds varying from 0.5 to 2 m/s, and a depth of operation varying from 200 to 6000 m. The length of AUVs varies from 1.42 m (AUV Cormoran) to 10 m (AUV Urashima). They mostly have torpedo shaped hull forms. AUVs can be employed to collect data samples of physical characteristics of water such as temperature, salinity, density, depth and conductivity and map out hydrothermal vents, tsunamis etc. AUGs are underwater robots which can hold their positions by gliding against the current or waves by making themselves neutrally buoyant and drift with the currents and waves or rest on the sea bed. They do not require thrusters or propellers for propulsion. AUGs are capable of carrying out a mission economically in comparison to AUVs and have much larger endurance. Few existing AUGs even have the capability to derive their propulsive energy from the ocean itself. One concept is to use the temperature differences in the ocean thermocline (principle of thermal stratification to convert heat into mechanical energy). In general, these vehicles are very small in size. Their operating speed varies from 0.1 to 0.5 m/s.

Towed fish are torpedo shaped bodies without any active propulsion. They are towed by a ship by a mooring line and have limitations on the depth of operation. Their depth of operation is usually limited to about 200 m. Typically, they carry instruments such as acoustic Doppler current profiler, DRAKE (Depth and Roll Adjustable Kite for Energy flux measurements) etc. During their deployment, the ship speed is typically in the range of 2.5–5.5 m/s. ROVs are tethered vehicles whose mission plan is executed from onboard a ship. They are often linked to the ship by either a neutrally buoyant tether or a load carrying umbilical cable. Their depth of operation varies from 200 to 11,000 m. They can be employed in exploration and mapping of seabed minerals.

Unlike ROVs, deep submergence vehicles (DSV), autonomous surface vehicles (ASV) and AUGs are unmanned submersibles without tethers or umbilical cables and follow a predefined path without operator intervention. They usually carry the power source onboard in the form of batteries and have a payload in accordance with their mission. The power supply unit and the speed details of few existing AUVs are given in **Table 1**.

Even though the initial interest in AUVs was developed for oceanographic research in the late 1960s by the University of Washington using Special Purpose Underwater Research Vehicle

AUV	Speed (knots)	Endurance (km or h)	Energy storage/supply device
ARCS (Canada)	5.5 (max.) 4 (cruising)	36 km (1 Ni-Cd battery) 72 km—2 Ni-Cd 235 km—Al-O_2	1 or 2 Ni-Cd battery (10 kW-h. each) or Al-O_2 fuel cells (100 kW-h)
AURORA (Canada)	3.5 (max.) 1.5-2 (Cruising)	750 km	Li ion battery units
THESEUS (Canada)	4 (Cruising)	780 km	Ag-Zn battery units (360 kW-h)
HUGIN 3000 (Norway)	4 (Cruising)	6–8 h—Ni-Cd 36 h—Al-O_2	Ni-Cd battery units (3 kW-h) Al-O_2 semi-fuel cells (18 kW-h)
DeepC (Germany)	6 (max.) 0.5-4 (Cruising)	400 km 60 h	PEM fuel cells
AUTOSUB (UK)	2-4 (Cruising)	500 km 144 h	Alkaline primary battery units
OKPO 6000 (Korea)	3 (max.)	10 h	Ag - Zn battery units
MARIDAN 600 (Denmark)	4 (max.) 3 (Cruising)	36 km 10 h	Pb-H_2SO_4 battery units
CETUS (US)	5 (max.) 1.5-2.5 (Cruising)	20-40 km	Pb-H_2SO_4 battery units

AUV	Speed (knots)	Endurance (km or h)	Energy storage/supply device
REMUS (US)	5 (max.) 3 (Cruising)	46.3 km	Pb-H_2SO_4 battery units (400 kW-h)
AQUA EXPLORER 2 (Japan)	2 (max.) 1 (Cruising)	24 hrs	Li primary battery units (3870 W-h)
R-One Robot (Japan)	3.6 (max.) 2 (Cruising)	120 km 25 hrs	CCDE (60 kW-h)

Table 1. General characteristics of AUVs around the world (from Thomas, 2003).

(SPURV) (see **Figure 1(a)**), it is only recently AUVs are being actually used in oceanographic research. Few recent instances where AUVs and ROVs are employed are: the Autonomous Benthic Explorer (ABE) (see **Figure 1(b)**) deployed in 1996-1997 in the Juan de Fuca region off the coast of Oregon to map the magnetic characteristics of a new lava flow; Odyssey IIB (see **Figure 1(c)**) class vehicles from MIT employed to study the convective overturning associated with the mixing of fresh and salt water in the Haro Strait on the US-Canadian northwest border; Woods Hole Oceanographic Institute's Argo ROV (see **Figure 1(d)**) played a major role in the 1985 discovery of the wreck of the RMS Titanic and German battleship Bismarck; and OKPO-6000 (see **Figure 1(e)**) explored successfully 2300 m deep seabed in 1996 near the Dok-Do island in the East Sea of Korea. Few other examples of the present day AUVs are Japan's AUV Urashima (see **Figure 1(f)**), which is a torpedo-shaped vehicle with depth rating up to 3500 m and whose hull frames are made of titanium; Norway's HUGIN 3000 (see **Figure 1(g)**) is a tail boomed vehicle that can operate in 3000 m depth; Canadian AUV Theseus (see **Figure 1(h)**), which is a torpedo-shaped vehicle of length 10.7 m, diameter 0.127 m and depth of operation of to 1000 m; UK's AUV Autosub (see **Figure 1(i)**), which is also a torpedo-shaped vehicle with depth rating of 6000 m, endurance of 4400 hrs and with a hull made of titanium; Denmark's Atlas Maridan Seaotter Mk II (see **Figure 1(j)**), which is a modular flat fish design with a length of 3.65 m and depth of operation 600 m; AUG Seaglider (see **Figure 1(k)**) which is a tear drop shaped vehicle of length 1.8–2 m depending on the configuration and Indian AUV Maya (see **Figure 1(l)**) which is a torpedo-shaped vehicle that is 1.72 m long and 0.234 m in diameter.

1.1. DESIGN OF UVs

The design of UV is mission specific and each UV is unique in design because it needs to cater to its unique set of mission requirements. However, the design objectives related to their underwater usage are based on hydrodynamic drag, power, propulsion, maneuvering and buoyancy control. Of these, the hydrodynamic drag is most important because it directly affects the power requirement, range, and endurance. Therefore, minimization of drag is a central objective in AUV design and it is an important problem in the area of marine hydrodynamics. This can be accomplished, in general, by some combination of (i) streamlined shaping of the hull, (ii) controlling boundary-layer, e.g., polymer injection or slot suction, (iii) energy-saving propulsion; e.g., a wake adapted propeller or a suction slot with a stern jet, and (iv) efficient maneuvering consistent with hydrodynamic stability. The first two of this list

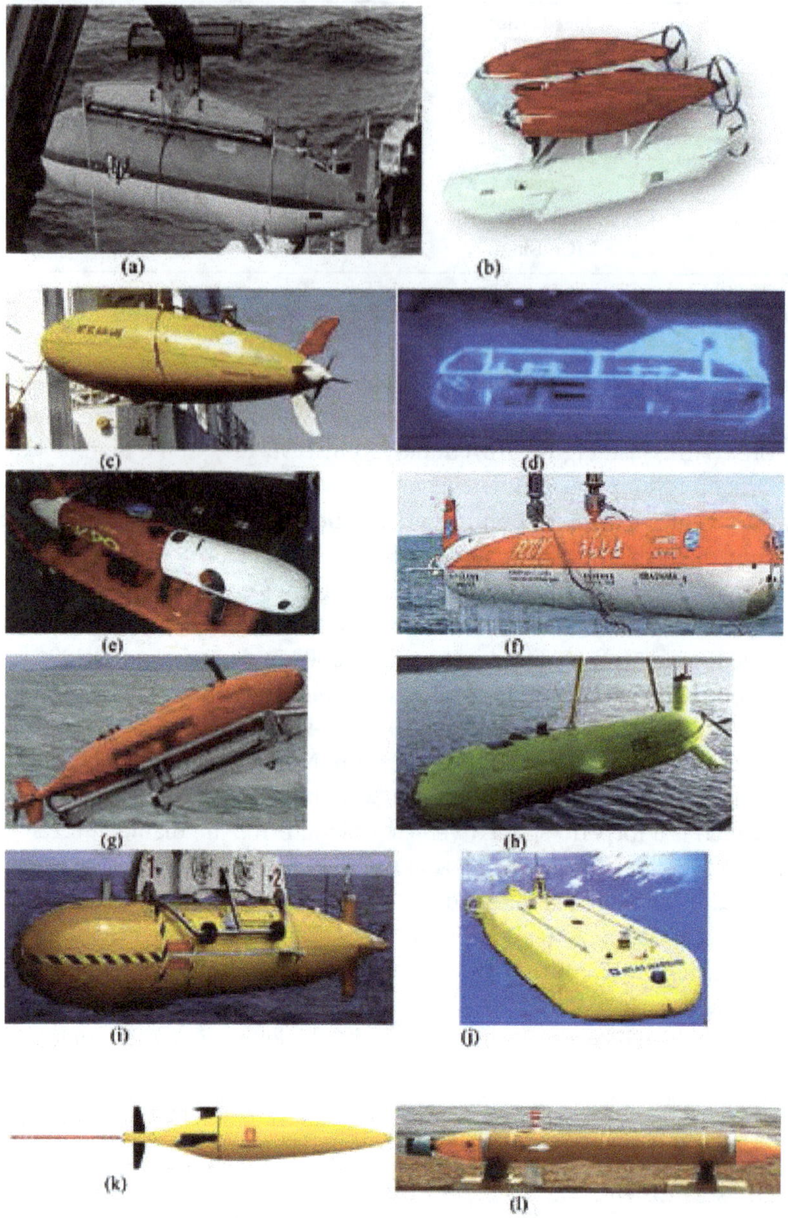

Figure 1. Diverse forms of AUVs. (a) AUV SPURV [2], (b) Autonomous Benthic Explorer [3], (c) AUV Odyssey IIB [4], (d) ROV Argo [5], (e) AUV OKPO 6000 [6], (f) AUV Urashima [7], (g) AUV Hugin 3000 [8], (h) AUV Theseus [9], (i) AUV Autosub [10], (j) AUV Atlas Maridan 300 [11], (k) AUG Seaglider [12], (l) AUV Maya [13].

attempt to reduce skin friction and pressure drag while the third attempts to extract energy lost to the fluid surrounding the vehicle. A complete systems design must simultaneously take into account all these four aspects though the complex nature of the complete problem does not permit analytical systems design approach [1].

For propelled vehicles, a body that has minimum drag need not be the one that requires minimum power because the benefit from the reduction in drag may be lost because of poor wake fraction which implies poor propulsive efficiency of the propeller which is fitted in the vehicle's wake. The efficiency of the propeller located in the wake of the hull is in general inversely proportional to the wake fraction and also to the thrust deduction factor. In order to optimize hull shapes of UVs, both drag (minimization) and wake fraction (minimization) should be considered simultaneously.

At present, the UV design process is mainly dominated by ad-hoc approaches that either use design experience or rely on simple rules of thumb [2] based on empirical formulations of hydrodynamic drag and wake fraction. Although an empirical approach is convenient at the preliminary design stage, it does not consider the local fairing effects on the flow, which play an important role in the estimation of drag and wake fraction. To overcome these limitations, the experimental approach using model testing in towing tank is often used, which, however, is both time consuming and expensive and, as a result, cannot be done for many designs. Typically, a maximum of three designs can be tested in a towing tank, which is certainly not enough for establishing a near optimum design. In this regard, the recent advances in CFD can play an important role in the UV design because with CFD the local fairing effects and the variations of flow can be accurately predicted so that hydrodynamic evaluations of many designs are possible in a short time economically. As a result, a near optimum design can be obtained in principle if CFD approach is integrated into the design process. Use of CFD to analyze the flow field around the hull and to perform computations of viscous drag has found interesting applications in ship design [2–4].

Although CFD has the advantage of reducing the time and cost of each analysis, it is difficult to manually change the design parameters of the UV hull form and conduct each analysis to obtain an optimized shape. Hence, there is a requirement to solve the problem with a process of optimization which is robust and automatic. Such attempts have been made for ships [3, 4] with limited success. In recent years, multidisciplinary design optimization (MDO) methods are being increasingly and effectively used to identify optimal designs [2].

The optimization of hull shape of UVs therefore, will ideally involve simultaneous minimization of drag and wake fraction and maximization of volume subject to constraints on the parameter space. The methodology presented in this thesis seeks to establish such a capability wherein the optimization technique based on genetic algorithm (GA) and CFD solver are seamlessly integrated with the computer aided geometric design (CAGD) tool in a single code that requires no user intervention during the entire optimization process.

1.2. Literature review

A brief review of the literature is described in this section in chronological order.

Ref. [5] conducted experiments on a systematic series of 24 mathematically related streamlined bodies of revolution to measure drag force, Reynolds stresses, pressure, kinetic energy, axial and radial velocity profiles, stern and far wake at deep submergence. The body geometric parameters are fineness ratio, prismatic coefficient, nose radius, tail radius and the position of the maximum section. The mathematically derived series of bodies of revolution, which was designated series 58, are used to form the offsets of the models using a sixth degree polynomial.

Ref. [6] conducted experiments on a 1:6 spheroid (MS) of 1.578 m long and 25 cm maximum diameter in a wind tunnel at a speed of 12 m/s and measured pressure distributions, mean velocity profiles, and Reynolds stresses in the thick axisymmetric turbulent boundary layer near the tail of the body. The thick boundary layer is characterized by significant variations in static pressure across it and very low level of turbulence. It is concluded that the static pressure variation is associated with a strong interaction between the boundary layer and the potential flow outside it, while the changes in the turbulence structure appear to be a consequence of the transverse surface curvature. The main conclusion was that by using the thin boundary layer calculation it is not possible to predict the behavior of the flow in the tail region of a body.

Ref. [1] introduced an automatic synthesis approach to minimum drag axisymmetric hull shapes in non-separating flow with constraints on volume and speed. The optimization was done in a finitely constrained parameter space with eight-parameter 'rounded-nose- tail boom' bodies. The parameters are the zero radius of curvature of the nose, maximum diameter and its axial location, curvatures at the locations of the maximum diameter and aft inflection point, radius and slope at the aft inflection point and the terminal radius of the profile. The drag was evaluated exploiting laminar flow by avoiding flow separation. Ref. [7] developed a multi-criteria optimization model for ship design which is a problem with multiple local optima with widely varying sets of designs. Simulated annealing had been successfully implemented to optimize the ship design process as a global optimization tool. The decision system was based on the analytic hierarchy process.

Ref. [8] discussed the Approximation Management Framework (AMF), for solving optimization problems which aims to use the cheaper low fidelity models in iterative procedures with occasional, but systematic, recourse to expensive high fidelity models. Three versions of AMF using nonlinear programming algorithms are implemented on a three dimensional (3D) aerodynamic wing and a two dimensional (2D) airfoil. The three methods discussed were augmented Lagrangian AMF, sequential quadratic programming AMF and AMF based on 'multilevel algorithm for large scale constrained trust-region optimization'.

Ref. [9] worked on implementing a downhill simplex method with constraints on volume displacement and transverse moment of area of water-plane of a ship to minimize its total drag. The parent hull form considered is classical Wigley hull of length 122 m defined by NURBS (Non-Uniform Rational B-Spline) surface. The ITTC formula is used to calculate the frictional drag and the wave drag was predicted using zeroth-order slender-ship approximation. Ref. [3] developed two simulation based ship design approaches, one of which is 'narrow band derivative-free' approach and the other is 'variable fidelity' approach. They used CFD to evaluate the objective function, which is the total drag of a ship, and validated the results by conducting model tests. Similar work was done by [10, 11].

Ref. [12] studied practical and quantitative methods for measuring effectiveness in naval ship design. A method is presented that uses the analytic hierarchy process combined with multi-attribute value theory to build an overall measure of 'effectiveness' and 'overall measure of risk function' using trained expert opinion to replace complex analysis tools. Ref. [13] implemented topological optimization techniques to reduce the weight of the composite advanced sail structure. The approach is applied to reinforcement layout optimization under an asymmetric wave slap loading condition. A high complexity model in the form of multilayered shell and a low complexity model in the form of stiffened shell are developed for layout optimization. Ref. [14] developed a framework for design optimization for problems that involve two or more objectives which may be conflicting in nature. The framework is implemented for the design of space propulsion involving a response surface based multi-objective optimization of a radial turbine for a liquid rocket engine. The surrogate model is integrated with GA-based Pareto front construction and can be effective in supporting global sensitivity evaluations. It has been concluded that a global sensitivity analysis provided a summary of the effects of design variables on objective function analysis and it is determined that no variable could be eliminated from the analysis.

Ref. [15] developed an underwater glider ALEX with independently controllable main wings and conducted wind tunnel experiments. To establish a mathematical model of the glider CFD is used to estimate the hydrodynamic forces acting on it and the results were compared with the experiments. Ref. [16] implemented a simulated annealing technique for the shape optimization of Cormoran AUV operating at snorkeling depths with constraints on surface area and volume and validated the results by conducting experiments. The objective was to minimize the wave making resistance at snorkeling depths while using an empirical formula for the calculation of the viscous drag. Ref. [17] developed a design optimization process for AUV using GA with cost, effectiveness and risk as the main design objectives. The design parameters are diameter, length to diameter ratio, forward shape coefficient, aft shape coefficient, endurance, speed, communication, payload, propulsion, battery and electronics configurations, wall thickness and material.

Ref. [18] studied the drag and turbulent noises created by the equipment like sonar array, electronic devices, antennas and video cameras. Automatic multi-objective optimization is applied using 'design of experience' as well as GA. Ref. [19] developed a Multidisciplinary Design Optimization (MDO) approach to the design of submarine considering four objectives, namely, deck area, drag, structural design and maneuvering. Particle Swarm Optimization (PSO) technique is used. The length of the parallel middle body, maximum diameter, tail shape parameter and nose shape parameter are the design variables. Ref. [20] developed an AUV PICASSO (Plankton Investigatory Collaborating Autonomous Survey System Operon). To improve the overall propulsive performance and maneuverability numerical and experimental study has been carried out and concluded that by changing the fore and aft hull form propulsive performance has been improved.

Ref. [21] developed a hybrid driven underwater glider Petrel. CFD simulations were carried out to study the glide mode of the glider. The glide efficiency is found to be significantly influenced by the chord length of the wing, the stability of the vehicle is influenced by the sweep angle of the wing and the glide stability is influenced by the location of the wing. Ref. [22] developed the Flinders AUV. Shape optimization of the vehicle with a ducted propeller has been carried out using CFD software. The Design of Experiments (DOE) method based single objective optimization problem with minimizing drag as the objective has been formulated and solved with location of the sail, the separation between sail and transponder and angle of attack of the nozzle as the variables. Ref. [23] developed an approach to characterize the spiraling motion of underwater gliders and applied on Seawing underwater glider. The hydrodynamic coefficients are computed using CFD. The proposed approach has been validated by conducting field experiments. Ref. [24] investigated hydrodynamic characteristics like drag and lift forces of the USM shallow underwater glider whose length is 1.3 m with 0.17 m diameter by using CFD.

2. Geometry definition of the hull forms

The geometry of the UV is axisymmetric and is shown in **Figure 2**. The parameterized shape of the body is given by [16]:

$$
\begin{aligned}
&r(x) = r_{\max}\left(1-\left(\frac{L_n-x}{L_n}\right)^{n_n}\right)^{1/n_n} \quad \text{for } 0 \le x \le L_n\\
&r(x) = r_{\max} \quad \text{for } L_n \le x \le L_n + L_m\\
&r(x) = r_{\max}\left(1-\left(\frac{x-L_n-L_m}{L_t}\right)^{n_t}\right) \text{ OR } r_{\max} + \frac{(x-L_n-L_m))(r_t-r_{\max})}{L_t} \quad \text{for } L_n + L_m \le x \le L\\
&L_n + L_m + L_t = L
\end{aligned}
\tag{1}
$$

where $r_{\max}$ is the maximum radius of the body so that the maximum diameter is d_{max} (=$2r_{max}$) and a middle body length L_m has this radius, $r\,(x)$ is the variation of radius over the total length

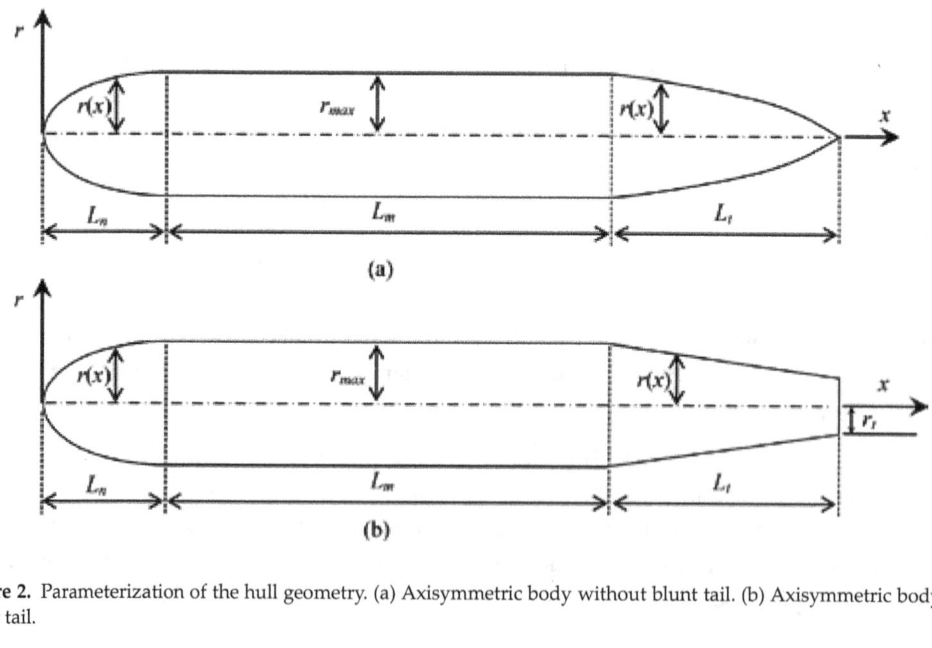

Figure 2. Parameterization of the hull geometry. (a) Axisymmetric body without blunt tail. (b) Axisymmetric body with blunt tail.

L, L_n and L_t are the lengths of the nose and the tail respectively and the exponents n_n and n_t are associated with the nose and tail shapes respectively.

The shape given by Eq. (1) leads to a conical nose shape (i.e. linear $r(x)$) for $n_n = 1$ and a conical tail shape for $n_t = 1$. For large values of n_n, the nose shape profile approaches a rectangle (i.e. $r(x)$ approaches r_{max}) and for large values of n_t, the tail shape profile also approaches a rectangle. For $n_n < 1$ and $n_t < 1$, the nose and tail shapes reverse the sign of their curvatures. The volume (∇), a design variable, is

$$\nabla = \pi \int_0^L r^2(x)dx \tag{2}$$

In this work, we have chosen two axisymmetric vehicles, namely, toy submarine USS Dallas (**Figure 3**) and AUV Cormoran (**Figure 4**). The parameters of these five vehicles are given in **Table 2**. It is to be noted that there is no parent hull form for the submarine. Its parameters, given in **Table 2**.

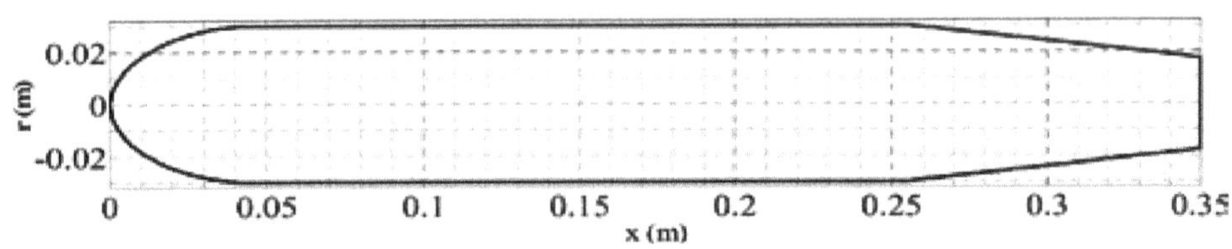

Figure 3. Toy submarine USS Dallas.

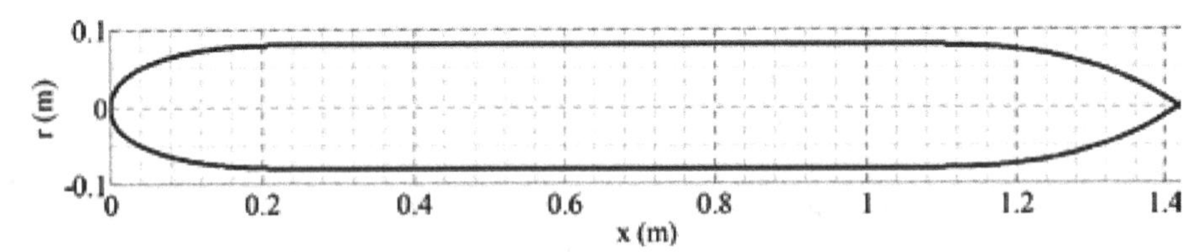

Figure 4. AUV Cormoran.

Parameters	Toy submarine USS Dallas	AUV Cormoran
Parameters of two hull shapes		
L_n (m)	0.045	0.24
L_m (m)	0.210	0.8
L_t (m)	0.095	0.38
L (m)	0.35	1.42
r_{max} (m)	0.03	0.08
r_t (m)	0.0175	0

Parameters	Toy submarine USS Dallas	AUV Cormoran
Parameters of two hull shapes		
n_n	1.9	2.3
n_t	1	3
l_s (m)	—	—
U (m/s)	0.5	1
S (m^2)	0.0612	0.63
∇(m^3)	0.000848	0.0245
$L/2r_{max}$	5.8	8.875
$\nabla^{2/3} / S$	0.6795	0.134

Table 2. The hull parameters.

Toy submarine USS Dallas was studied by [2] who reported drag force obtained from CFD simulation for a speed of 0.5 m/s which translates to $Re_L = 1.75 \times 10^5$. AUV Cormoran was studied by [16] who reported drag coefficients for a few speeds measured in towing tank.

3. Computational domain and discretization

The computational domain is shown in **Figure 5** where S_B is the body boundary, S_{F1} is a surface enclosing the subdomain Ω_1 as well as S_B and S_F is a surface enclosing the entire domain Ω, which encloses Ω_1. The domain Ω, bounded by S_F and marked ABCD in the x-z or the vertical plane (x-y or the horizontal plane being similar since the body is axisymmetric), has a length 6.2 L (L being the length of the body) and a height (i.e. breadth in the z direction) of 1.2 L. The domain length upstream is 0.7 L and the domain length downstream is 4.5 L. The domain is large enough to capture the entire viscous-inviscid interaction and the wake development. It should be noted that the domain Ω embeds in it the domain Ω_1 as well as the body surface S_B. The mesh of the domain Ω is modeled by a uniform grid such that on the boundaries AB and CD, the number of cells is N_z ($=N_y$) and on the boundaries AD and BC, the number of cells is N_x. Thus, the domain Ω has a grid of $N_x \times N_y \times N_z$ cells in x (length), y (breadth) and z (height) directions of the cuboid Ω.

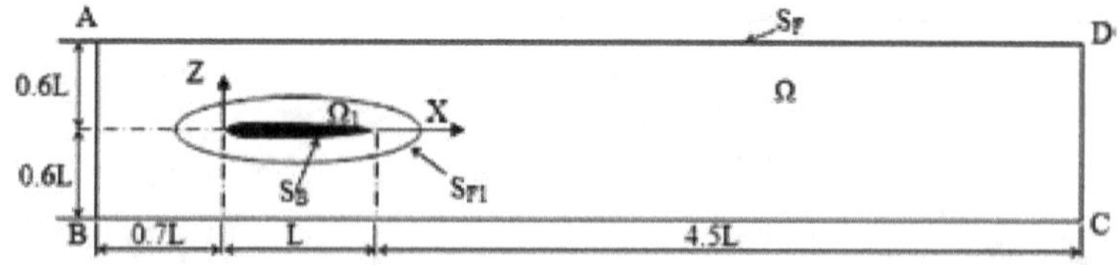

Figure 5. Fluid domain and boundaries for CFD calculations of AUV Cormoran, Afterbody 1 and AUV Autosub.

The subdomain Ω_1 is defined such that a point on S_{F1} is at a (perpendicular) distance of l from S_B and is meshed with N_r cells over this distance l with a graded mesh. The grid point nearest to S_B (wall adjacent cell size) is located at a (perpendicular) distance of l_1 from it and a successive ratio (g), defined as the ratio of successive distances between grid points normal to the body surface, is prescribed. This ensures that

$$l = \sum_{i=1}^{N_r} l_1 g^{i-1} \tag{3}$$

and the mesh over Ω_1 is an H-type structured mesh.

The successive or growth ratio (g) should be so chosen that it prevents the wall adjacent cells from being placed in the buffer layer of $y^+ = 1$. The acceptable distance between the cell centroid and the wall adjacent cells is usually measured in the wall unit y^+. However, it requires some trial and error to determine a suitable value of l_1. In all calculations, a value of $l_1 = 0.001$ mm was adopted which is found to satisfy the $y^+ < 1$ requirement. The mesh of the domain Ω_1, which embeds the body, was modeled by adopting $g = 1.1$ and a grid of $N_x \times N_r \times N_\theta$ cells in longitudinal, radial and circumferential direction respectively. The interface between Ω_1 and Ω is handled by the CFD solver by constructing interfacial cells.

3.1. Boundary conditions

The conditions imposed on boundaries of the fluid domain (see **Figure 5**) are:

a. At the inlet boundary AB, velocity U normal to the boundary is specified.

b. On the boundary CD, the pressure outlet boundary condition has been imposed. It implies that the pressure (p) is set to gauge pressure (i.e. $p = 0$) and the gradients of k and ω are set to zero.

c. On the top boundaries AD, BC ($z = \pm 0.6\ L$) and the two side boundaries ($y = \pm 0.6\ L$), the zero shear stress condition has been imposed.

d. On the surface boundary S_B, the no slip condition is specified.

3.2. Solver parameters

The commercial CFD solver has been used and the solver parameters are presented in **Table 3**. The velocity components are governed by the momentum equations. The Roe flux algorithm is used for coupling the pressure and velocity terms. Second order upwind scheme is adopted for the discretization of pressure, momentum, turbulent kinetic energy and turbulence dissipation rate. The convergence criterion of 10^{-4} is set for velocity components and 10^{-6} for continuity, k and ω. The termination of the program is based on the final steady value of drag. All simulations were run using 3D steady segregated RANS solver. In all the optimization problems treated in this thesis, the number of iterations was fixed at 2000, and the drag force has been obtained by its average over one cycle (between a peak and a trough) preceding the last iteration number.

Discretization	
Pressure	Body force weighted
Momentum; turbulent kinetic energy and dissipation rate	Second order upwind
Fluid properties	
Density (ρ)	1000 kg/m^3
Kinematic viscosity (ν)	10^{-6} m^2/s
Roughness parameters	
Roughness constant (C_S)	0.5
Roughness height (K_S)	0
Turbulence criteria at inlet	
Turbulent intensity (T)	5%
Turbulent length scale (l_T)	0.001 L
Under relaxation factors	
Pressure	0.3
Density	1
Body forces	1
Momentum	0.7
Turbulent viscosity	1
Turbulent kinetic energy	0.8
Turbulent dissipation rate	0.8

Table 3. Solver parameters and constants used in the study.

This reduces the computational effort significantly because instead of about 4000–6000 iterations for an accurately converged value of drag, one can use only 2000 iterations for all intermediate configurations during the optimization process. The drag of the final optimized configuration, however, is obtained by approximately 4000–6000 iterations for high accuracy.

4. CFD analysis of toy submarine USS Dallas

The mesh of the domain Ω (see **Figure 5**) for toy submarine USS Dallas has 260,100 cells ($=N_x \times N_y \times N_z = 100 \times 51 \times 51 = 260{,}100$) with $g = 1$, which gives a cell size of about 8.2 mm on the boundaries AB and CD, and a cell size of about 21.7 mm on the boundaries AD and BC. The mesh in the domain Ω_1 has 57,500 cells ($=N_x \times N_r \times N_\theta = 100 \times 115 \times 50 = 575{,}000$) with $g = 1.1$, which gives the distance between S_B and S_{F1} (along any one of the rays) as $l = 52$ mm with the size of the cell adjacent to S_{F1} as 4.3 mm and in the circumferential direction an angle of 7.2 deg. The length of wall adjacent cell (l_1) is 0.001 mm. The discretization of the domains Ω and Ω_1 are shown in **Figure 6** for the toy submarine.

The comparison of the drag forces obtained using CFD as reported by Alam *et al.* (2015) with present CFD calculations are recorded in **Table 4** showing very good match (within 0.2% difference). The drag values obtained by Eqns. (2.3), (2.4) and (2.5) were reported by Alam

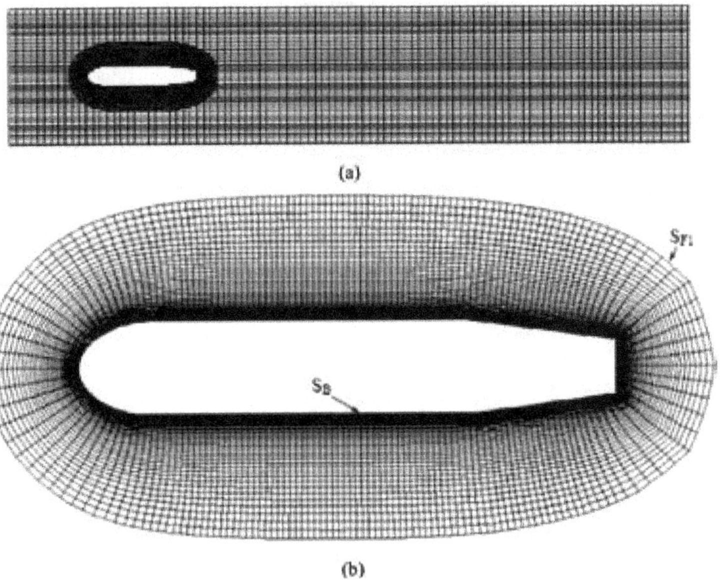

Figure 6. Mesh for toy submarine USS Dallas (No. of nodes 835,100; no. of cells 956,664). (a) Mesh in the full domain for toy submarine USS Dallas. (b) Mesh in the Ω_1 for toy submarine USS Dallas.

Source	D (N)
CFD (Alam *et al.* (2015) (realizable k-ε model)	0.1065
CFD (Present) (SST k-ω model)	0.1062
Difference (%)	−0.2

Table 4. Comparison of drag forces of toy submarine USS Dallas at U = 0.5 m/s (Re_L = 1.75 × 10^5).

et al. (2015). The mesh convergence study is not reported here because the present mesh with about 1 million cells is much finer than 0.69 million cells used in Alam *et al.* (2015).

4.1. CFD analysis of AUV Cormoran

The mesh of the domain Ω (see **Figure 5**) for AUV Cormoran has 26,010 cells (= $N_x \times N_y \times N_z$ = 100 × 51 × 51 = 260,100) with g = 1, which gives a cell size of about 33 mm on the boundaries AB and CD, and a cell size of about 92.3 mm on the boundaries AD and BC. The mesh in the domain Ω_1 has 575,000 cells (= $N_x \times N_r \times N_\theta$ = 100 × 115 × 50 = 575,000) with g = 1.1, which gives the distance between S_B and S_{F1} (along any one of the rays) as l = 52 mm with the size of the cell adjacent to S_{F1} as 4.3 mm and in the circumferential direction an angle of 7.2 deg. The length of wall adjacent cell (l_1) is 0.001 mm. The discretization of the domains Ω and Ω_1 are shown in **Figure 7**.

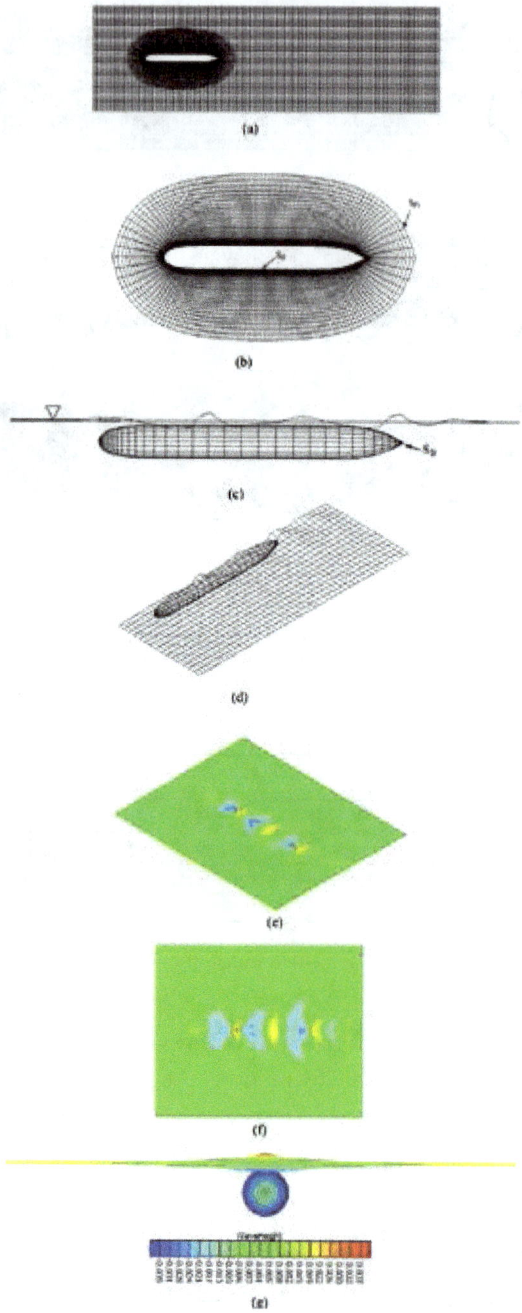

Figure 7. Mesh for AUV Cormoran. (a) Mesh in the full domain for AUV Cormoran(No. of nodes 835,100; no. of cells 956,664). (b) Mesh in the Ω_1 for AUV Cormoran (c) panels on AUV Cormoran. (d) Panels on AUV Cormoran and free surface (one half shown) (No. of panels = 3500 on S_{B}, = 1500 on free surface, = 5000 total). (e) Surface wave pattern generated by AUV Cormoran (U = 1 m/s). (f) Surface wave pattern generated by AUV Cormoran in top view (U = 1 m/s). (g) Surface wave pattern generated by AUV Cormoran in front view (U = 1 m/s).

(a) Comparison with experiments				
U (m/s)	C_D	C_W	$C_T = C_D + C_W$	
	CFD (Present)		CFD (Present)	Experimental (Alvarez *et al.* [2008])
0.916	0.004524	0.004524	0.009048	0.00913
1.062	0.004463	0.004866	0.00932	0.00989
1.195	0.004325	0.007827	0.0121	0.01291
1.399	0.004222	0.003998	0.00822	0.00871
(b) Comparison with empirical formulae				
U (m/s)	C_D			
	CFD			
0.916	0.004524	0.0047	0.004692	0.004729
1.062	0.004463	0.004555	0.00455	0.004585
1.195	0.004325	0.00445	0.00444	0.004474
1.399	0.004222	0.004309	0.0043	0.004333

C_W, coefficient of wave resistance.

Table 5. Comparison of drag coefficients of AUV Cormoran.

The experimental drag forces as reported by Alvarez *et al.* (2008) are for snorkeling depth of submergence ($z = -0.05$ m, see **Figure 5**) which will induce wave making drag in addition to viscous drag. In order to capture the wave-making component of drag, a free surface boundary condition is specified on the boundary AD (see **Figure 5** and **7b**). The dimensions of the free surface are 2 L in the x-direction and L in the y-direction. The surface is discretized with 50 equispaced panels in the x-direction, which gives a panel size of 28.4 mm and with 30 equispaced panels in the y-direction, which gives a panel size of 35.5 mm, making a total of 1500 panels. On the surface S_B, in the x-direction, for $0 \le x \le L_n$ number of panels are 40 with a g = 1.1, in between $L_n \le x \le L_n + L_m$, 10 with g = 1 and between $L_n + L_m \le x \le L$, 20 with g = 1.1 making a total of 70 panels and in the circumferential direction, 50 equispaced panels, each making an angle of 7.2 deg. The surface wave pattern generated by AUV Cormoran for $U = 1$ m/s is shown in **Figures 7(e), 7(f)** and **7(g)**.

The comparisons of the drag forces are recorded in **Table 5** for four forward speeds. As can be seen, the present calculations are very accurate.

5. Conclusion

A critical review of the literature brings out the following: Several AUV, AUG, submarine and similar vehicle shapes had been treated using the CFD approach. A nearly exhaustive list of these shapes are: Afterbody 1, Afterbody 2, modified spheroid and F57, AUV Rainbow, AUV Cormoran, AUV Autosub and AUV Soton, AUG Alex, AUV PICASSO, AUV Petrel, AUG USM, AUV

Flinders, toy submarine USS Dallas and AUG Seawing. From the literature, it is observed that the use of CFD in the hydrodynamic design of UVs is finding increasing acceptance, as is evident by the fact that about 15 UV shapes have been treated. The CFD methodology adopted in the present work has been validated with a few example problems from the literature. The hydrodynamic parameters computed by CFD are drag and wave resistance.

Author details

Kondeti Lakshmi Vasudev

Address all correspondence to: kondetilv@hindustanuniv.ac.in

Department of Mechanical Engineering, Hindustan Institute of Technology and Science, Chennai, India

References

[1] Parsons JS, Goodson RE, Goldschmied FR. Shaping of axisymmetric bodies for minimum drag in incompressible flow. Journal of Hydronautics. 1974;**8**(3):100-107

[2] http://www.navaldrones.com/SPURV.html

[3] http://auvac.org/configurations/view/135

[4] http://auvlab.mit.edu/MURI/1997_Rprtfinal.html

[5] http://www.whoi.edu/page.do?pid=83577&tid=3622&ci

[6] http://www.geojetimes.co.kr/news/articleView.html?idxno=25928

[7] http://www.nauticexpo.com/prod/mitsubishi–heavy–industries–ship–ocean

[8] http://www.act–us.info

[9] http://www.ise.bc.ca/theseus.html

[10] http://www.unmanned.co.uk/platforms/unmanned–vehicles–news/page/108/

[11] http://www.act–us.info/sensor_list.php?cat=AUV&type=Sensor%20Type

[12] http://www.km.kongsberg.com/ks/web/nokbg0240.nsf/AllWeb/EC2FF8B58CA491A4C1257B870048C78C?OpenDocument

[13] http://www.nio.org/index/option/com_newsdisplay/task/view/tid/4/sid/23/nid/88

[14] Alam K, Ray T, Anavatti SG. A new robust design optimization approach for unmanned underwater vehicle design. Journal of Engineering for the Maritime Environment. 2012;**226**(3):235-249

[15] Campana EF, Peri D, Tahara Y, Stern F. Shape optimization in ship hydrodynamics using computational fluid dynamics. Computer Methods in Applied Mechanics and Engineering. 2006;**196**(1-3):634-651

[16] Shahid M, Huang D. Computational fluid dynamics based bulbous bow optimization using a genetic algorithm. Journal of Marine Science and Applications. 2012;**11**:286-294

[17] Gertler M. Resistance experiments on a systematic series of streamlined bodies of revolution—For application to the design of high-speed submarines. In: DTRC Report: C-849. Washington, D.C; 1950

[18] Patel VC, Nakayama A, Damain R. Measurements in the thick axisymmetric boundary layer near the tail of a body of revolution. Journal of Fluid Mechanics. 1974;**63**(2):345-367

[19] Ray T, Gokarn RP, Sha OP. A global optimization model for ship design. Computers in Industry. 1995;**26**:175-192

[20] Alexandrov N, Dennis JE Jr, Lewis RM, Torczon V. A trust region framework for managing the use of approximation models in optimization. Structural Optimization. 1998;**15**(1): 16-23

[21] Percival S, Hendrix D, Noblesse F. Hydrodynamic optimization of ship hull forms. Applied Ocean research. 2001;**23**(6):337-355

[22] Peri D, Campana EF. High fidelity models and multi-objective global optimization algorithms. Journal of Ship Research. 2005;**49**(3):159-175

[23] Tahara Y, Tohyama S, Katsui T. CFD based multi-objective optimization method for ship design. International Journal for Numerical Methods in Fluids. 2006;**52**:499-527

[24] Demko D. Tools for Multi-Objective and Multi-Disciplinary Optimization in Naval Ship Design [Thesis]. Virginia Polytechnic Institute and State University; 2006

8

Lunar Science: Internet for Space Tourism

Ayodele Abiola Periola

Abstract

The increased interest in space exploration drives the development of novel technologies that are useful in other areas, such as aviation. The use of these technologies gives rise to new challenges and applications. Space tourism is an emerging application due to advances in space exploration technologies. This paper addresses two challenges aimed at ensuring continued internet access in space tourism. The first is designing network architecture to ensure continued internet access for space tourists aboard a space vehicle. The second is using aerial vehicle technology to enhance access to cloud content in areas with poor telecommunication infrastructure. The paper proposes the distributed handover algorithm ensuring that the space vehicle can execute handover from terrestrial wireless networks to aerial platforms and satellites as a last mile connection. It also proposes the concept of aerial diversity ensuring low cost access to cloud content. Performance simulation shows that the use of the distributed handover algorithm enhances channel capacity by 18.4% on average and reduces latency by 11.6% on average. The use of the cloud content access system incorporating aerial diversity enhances the channel capacity of terrestrial wireless networks by up to 85% on average.

Keywords: Space Tourism, Wireless Communications, Wireless Handover, aerial platforms, satellites, space tourist

1. Introduction

The internet comprises multiple converging technologies that interact together in a global network. Information access via the internet faces a significant number of challenges. These challenges influence the ease with which information can be accessed via the internet. The quality of service (QoS) associated with internet access is determined by metrics such as channel capacity, latency, throughput and packet loss rate.

Advances in networking have played a significant role in internet evolution. For example, the internet initially used wired technology as the communication media; however, the internet is now accessed via wireless radio [1–3]. This transition increases the mobility of subscribers seeking to access data [4, 5]. The emergence of smartphones has improved subscriber ability to access the internet. This increased access requires network algorithms to support the realization of enhanced QoS in fifth generation (5G) wireless networks and beyond 5G (B5G) networks.

Internet access via wireless technologies benefits from new technologies such as: (i) new variants of the internet protocol (IP) and the transmission control protocol (TCP) [6–10], (ii) improved packet switching [11–13], (iii) World Wide Web [14], (iv) IEEE 802.11 wireless network standard [15], and (v) artificial intelligence [16, 17].

Currently, there is increased interest in space exploration leading to the development of technologies such as small satellites [18, 19] and aerial vehicles such as stratospheric platforms [19] and drones [20]. The development of these technologies enables capital constrained organizations to engage in space exploration. This also enables the emergence of new applications requiring internet access such as space tourism. The emergence of space tourism [21–39] requires a solution to providing uninterrupted internet access to subscribers aboard a space vehicle, as well as improving accessibility to the cloud content internet.

S/N	Acronym	Meaning
1	APSH	Aerial Platform to Satellite Handover
2	C-RAN	Cloud Radio Access Network
3	DHA	Distributed Handover Algorithm
4	eNB	Evolved Node B
5	gNB	Next generation Node B
6	MAV	Manned aerial vehicle
7	MIPv6	Mobile internet protocol version 6
8	P-GW	Packet data gateway
9	PMIPv6	Proxy mobile internet protocol version 6
10	QoS	Quality of Service
11	S-GW	Serving gateway
12	SMIPv6	Seamless mobile internet protocol version 6
13	SISH	Sub –orbital Intersatellite Handover
14	TCP	Transmission Control Protocol
15	TWAH	Terrestrial Wireless Network to Aerial Network Handover
16	TWNH	Terrestrial Wireless Network Handover
17	UAV	Unmanned Aerial Vehicle

Table 1. Acronyms used in this paper.

S/N	Parameter	Meaning
1	N	Set of wireless networks
2	C	Set of cloud platforms
3	S	Set of subscribers
4	n_S	Set of satellite networks
5	n_T	Set of terrestrial wireless networks
6	n_S^r	The $(r)^{th}$ satellite network.
7	n_T^u	The uth terrestrial wireless network
8	C_i	The $(i)^{th}$ cloud platform hosting content that subscriber s_z seeks to access
9	s_z	The $(z)^{th}$ subscriber desiring access to cloud based content
10	$\alpha(n_T^u)$	The coverage region of n_T^u
11	$\alpha(n_S^r)$	The coverage region of n_S^r
12	β	Is the network sub – indicator
13	$I(\beta, C_i, t_j)$	Cloud access indicator at epoch t_j
14	$\emptyset$.	Null set
15	θ	Set of possible subscriber locations.
16	θ_g	Set of ground locations for subscribers
17	θ_{ae}	Set of aerial locations for subscribers
18	θ_g^c	The cth ground location
19	θ_{ae}^n	The nth aerial location
20	N'	Updated set of wireless networks
21	ϕ_{SU}	the network designed to provide access to cloud content for θ_{su}
22	$\mathbb{E}(s_x)$	Mean latency for subscriber s_x
23	$Th(s_x, \beta, t_j)$	Throughput associated with data accessed by $s_x; s_x \in S$ via network entity β at epoch t_j
24	$D(s_x, t_j)$	Size of data accessed by s_x from network entity β at epoch t_j
25	$P_{th}(\gamma)$	Threshold signal strength for terrestrial wireless network base station entity.
26	l_{th}	Threshold Latency
27	$\mathbb{E}_1(s_z)$	The mean latency computed for multiple subscribers.
28	γ	Set of terrestrial wireless network base station entities
29	Ƀ	Set of stratospheric platforms
30	p	Set of satellites
31	$\varkappa \in \{\gamma_b, Ƀ_l, p_t\}$	Entity $\varkappa$ denotes transmitting nodes in terrestrial, stratosphere and outer space respectively.
32	$P(\varkappa, t_j)$;	The strength of the signal form entity $\varkappa$ at epoch t_j
33	C_{ter}	Channel capacity of space vehicle in terrestrial plane
34	C_{ae}	Channel capacity of space vehicle in aerial plane
35	C_{sp}	Channel capacity of space vehicle in space plane.

S/N	Parameter	Meaning
36	$B(z', b)$	Bandwidth of channel z' for the bth terrestrial wireless network base station entity
37	$B(z', l)$	Bandwidth of channel z' for the lth stratospheric platform base station entity
38	$B(z', c)$	Bandwidth of channel z' for the cth in – orbit satellite.
39	$P_{tr}(\gamma_d, z')$	Transmit power between the space vehicle and terrestrial wireless network γ_d on channel z'
40	$P_{tr}(Þ_l, z')$	Data transmit power between the space vehicle and high altitude platform $Þ_l$ on channel z'
41	$P_{tr}(p_c, z')$	Data transmit power between the space vehicle and in – orbit satellite p_c on channel z'
42	$P_{int}(\gamma_d, z')$	Interference power between the space vehicle and terrestrial wireless network on channel z'
43	$P_{int}(Þ_l, z')$	Interference power between the space vehicle and high altitude platform on channel z'
44	$P_{int}(p_c, z')$	Interference power between the space vehicle and in – orbit satellite on channel z'
45	$h_{11}(\gamma_d, z')$	Transmit channel gain between space vehicle and terrestrial wireless network on channel z'
46	$h_{11}(Þ_l, z')$	Transmit channel gain between the space vehicle and high altitude platform on channel z'
47	$h_{11}(p_c, z')$	Transmit channel gain between the space vehicle and in – orbit satellite on channel z'
48	$h_{12}(\gamma_d, z')$	Interference channel gain between space vehicle and terrestrial wireless network on channel z'
49	$h_{12}(Þ_l, z')$	Interference channel gain between the space vehicle and high altitude platform on channel z'
50	$h_{12}(p_c, z')$	Interference channel gain between the space vehicle and in – orbit satellite on channel z'
51	C_{ave}	Average channel capacity for the space vehicle
52	$I(\gamma_d, Þ_l)$	Handover indicator between γ_d and $Þ_l$
53	$I(Þ_l, p_c)$	Handover indicator between $Þ_l$ and p_c
54	D'	Amount of transmitted data in bytes
55	β_1	Latency associated with data transmission in absence of proposed handover mechanism
56	β_2	Latency associated with data transmission after incorporating the handover mechanism
57	$P_{co}(\gamma_b)$	Probability of network congestion occurring on terrestrial wireless network γ_b
58	$Th(\gamma_b)$	Channel capacity of γ_b is denoted $Th(\gamma_b)$
59	$Th_1^{cl}(\gamma_b)$	Aggregate channel capacity of accessing cloud content without proposed cyber – physical system
60	$Th_2^{cl}(\gamma_b)$	Aggregate channel capacity of accessing cloud content with proposed cyber – physical system

Table 2. Set of notations used in this paper.

This chapter addresses two challenges: it designs (i) a network infrastructure with associated mechanisms to ensure continued access for space tourist subscribers aboard a space vehicle and (ii) a solution to improve the cloud service accessibility in developing nations. The chapter makes the following contributions:

1. Firstly, it proposes a network architecture that incorporates the space tourist subscriber in commercial space flights. The space tourist subscriber requires access to cloud-based content and the proposed network architecture ensures that there is a continuity of access to cloud content at every tier via the proposed handover mechanism.

2. Secondly, it proposes a novel architecture that incorporates aerial diversity i.e. use of unmanned and manned aerial vehicles to achieve access to cloud content. This has the benefit of reducing congestion on in terrestrial wireless networks. The architecture uses manned and unmanned aerial and robotic entities for information delivery in the internet.
3. Thirdly, it formulates the performance metrics and benefits for the proposed mechanisms. These metrics are examined considering networks that do and do not incorporate the proposed mechanism. The metric is the aggregate throughput for a network comprising multiple base station entities.

The rest of the paper is structured as follows. Section 2 formulates the problem being addressed in this chapter. Section 3 presents the proposed mechanisms. Section 4 formulates the performance model. Section 5 presents and discusses the simulation results and performance benefits. Section 6 is the conclusion.

The list of acronyms and the set of notations used in this paper are shown in **Tables 1** and **2** respectively.

2. Problem formulation

The discussion here is divided into three parts. The first part describes the system model. The second defines the problem and challenges being addressed in this chapter. The third focuses on the challenge being addressed as regards access to cloud based services.

2.1. System model

The network scenario comprises cloud radio access networks (C-RANs). Each C-RAN comprises a base station entity such as the evolved Node B (eNB) or next generation Node B (gNB). The eNB or gNB is connected to a cloud platform that provides resources in the network control plane. The base station entity is connected to cloud platforms that host content being demanded by subscribers. The system model assumes that subscribers can access the network i.e. cloud content at the desired epoch. The network comprises terrestrial wireless and satellite network segments. A scenario showing the network is shown in **Figure 1**.

The scenario in **Figure 1** shows the connection between two eNBs (i.e. eNB1 and eNB2) with overlapping coverage. The first eNB i.e. eNB 1 is connected to the cloud platform being the closer of the two eNBs. The packet data gateway (P-GW) of eNB 1 interacts with the gateway entity at the cloud platform hosting the content being accessed. The eNBs can execute handover to support the migration of subscriber S_H. This is realized by the dynamics associated with serving gateway (S-GW) in seamless handover execution [40].

2.2 Problem definition: the 'space tourist' subscriber

The considered scenario comprises multiple networks enabling subscribers to access cloud-based content. Let N, S and C be the set of wireless networks, subscribers and cloud platforms, respectively. Such as:

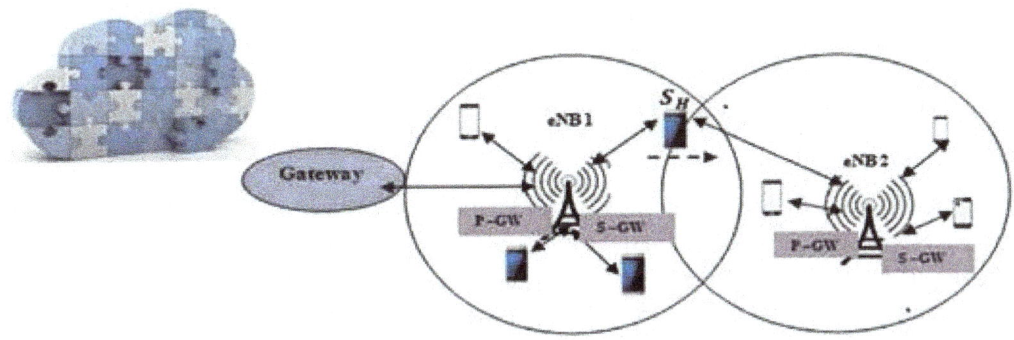

Figure 1. Network scenario showing the system model.

$$N = \{n_S, n_T\} \quad (1)$$

$$S = \{s_1, s_2, \ldots, s_x\} \quad (2)$$

$$C = \{C_1, C_2, \ldots, C_y\} \quad (3)$$

$$n_S = \{n_S^1, n_S^2, \ldots, n_S^p\} \quad (4)$$

$$n_T = \{n_T^1, n_T^2, \ldots, n_T^q\} \quad (5)$$

Where:

n_S and n_T are the set of satellite and terrestrial wireless networks respectively.

$s_z, s_z \epsilon\, S$ is the $(z)^{th}$ subscriber desiring access to cloud-based content.

$n_S^r, n_S^r \epsilon\, n_S$ is the $(r)^{th}$ satellite network.

$n_T^u, n_T^u \epsilon\, n_T$ is the $(u)^{th}$ terrestrial wireless network.

$C_i, C_i \epsilon\, C$ is the $(i)^{th}$ cloud platform hosting content that subscriber s_z seeks to access.

The coverage region of n_T^u and n_S^r are denoted as $\alpha(n_T^u)$ and $\alpha(n_S^r)$, respectively. Let $I(\beta, C_i, t_j)\ \epsilon\ \{0,1\}, \beta \epsilon \{ n_T^u, n_S^r \}, t_j\ \epsilon\ t, t = \{t_1, t_2, \ldots, t_w\}$ be the cloud access indicator at epoch t_j. The states $I(\beta = n_T^u, C_i, t_j) = 0$ and $I(\beta = n_T^u, C_i, t_j) = 1$ signify that the $(i)^{th}$ cloud platform C_i is inaccessible and accessible to base station entities of the $(u)^{th}$ terrestrial wireless network at epoch t_j respectively. The indicator $I(\beta = n_S^r, C_i, t_j) = 0$ and $I(\beta = n_S^r, C_i, t_j) = 1$ signify that the $(i)^{th}$ cloud platform C_i is inaccessible and accessible to base station entities of the $(r)^{th}$ satellite network, respectively. The ground-based entity of the $(r)^{th}$ satellite network is the terrestrial component of a satellite network.

A scenario described by the transition $I(\beta = n_T^u, C_i, t_j) = 0, I(\beta = n_T^u, C_i, t_{j+1}) = 1, t_{j+1} \epsilon\, t$ is one in which the cloud platform C_i is connected to network n_T^u at epoch t_{j+1} and not connected at epoch t_j respectively. Another plausible scenario is $(\beta = n_T^u, C_i, t_j) = 0$, $I(\beta = n_S^r, C_i, t_{j+1}) = 0$, $I\left(\beta = n_S^r, C_i, t_{j+j'}\right) = 0, t_{j+j'} \epsilon\, t$, which describes a case where subscriber s_z moves through the regions where access to cloud content via terrestrial network is infeasible at epochs t_j and t_{j+1} but feasible at epoch $t_{j+j'}$.

The variable $I(\beta, C_i, t_j)$ can have a varying number of contexts described by transitions between different scenarios for different β, C_i and t_j. A common factor across these scenarios is the implied assumption that $N \cap \{n_S \cup n_T\} \neq \emptyset$. However, this does not consider the requirement to provide internet access in outer–space. Hence, another scenario that is yet to be considered is one described as $N \cap \{n_S \cup n_T\} = \emptyset$ that considers the space tourist subscriber which has not been considered.

In the terrestrial plane, the subscriber s_z accesses the cloud content from a terrestrial location. However, cloud content can be accessed from other locations such as the ocean, and near space regions. Let θ denote the set of possible subscriber locations, such as:

$$\theta = \{\theta_g, \theta_{ae}, \theta_{su}\} \tag{6}$$

$$\theta_g = \left\{\theta_g^1, \theta_g^2, \ldots, \theta_g^v\right\} \tag{7}$$

$$\theta_{ae} = \left\{\theta_{ae}^1, \theta_{ae}^2, \ldots, \theta_{ae}^m\right\} \tag{8}$$

$$\theta_{su} = \left\{\theta_{su}^1, \theta_{su}^2, \ldots, \theta_{su}^f\right\} \tag{9}$$

Where

θ_g and θ_{ae} are the set of ground and aerial locations respectively.

$\theta_g^c, \theta_g^c \,\epsilon\, \theta_g$ is the cth ground location.

$\theta_{ae}^n, \theta_{ae}^n \,\epsilon\, \theta_{ae}$ is the nth aerial location.

θ_{su} is the set of locations in space.

$\theta_{su}^{v'}; \theta_{su}^{v'} \epsilon\, \theta_{su}$ is the v' sub – orbital location.

The definition of θ excludes the underwater and underground locations.

There is coverage for locations θ_g and θ_{ae} given that $I(\beta = n_T^u, C_i, t_j) = 1 \forall \theta_g, \theta_{ae}$ hold true. The condition $I(\beta, C_i, t_j) = 1 \forall \theta_g, \theta_{ae}$ indicates that there is no network coverage for locations θ_g and θ_{ae}. Satellite and terrestrial wireless networks cannot deliver cloud access to space tourist subscribers when:

$$\left\{I(\beta = n_T^u, C_i, t_j), I(\beta = n_T^u, C_i, t_{j+1}), I\left(\beta = n_T^u, C_i, t_{j+j'}\right), \ldots, I(\beta = n_T^u, C_i, t_w)\right\} = 0, \forall \theta_{su} \tag{10}$$

$$\left\{I(\beta = n_S^r, C_i, t_j), I(\beta = n_S^r, C_i, t_{j+1}), I\left(\beta = n_S^r, C_i, t_{j+j'}\right), \ldots, I(\beta = n_T^u, C_i, t_w)\right\} = 0, \forall \theta_{su} \tag{11}$$

This is because terrestrial and satellite networks do not provide internet access for space tourist subscribers.

Let N' denote the set of updated set of wireless networks such that:

$$N' = \{N, \phi_{SU}\} \tag{12}$$

Where ϕ_{SU} is the network designed to provide access to cloud content for θ_{su}, then it is desired that:

$$\left\{I((N' \cap N)', C_i, t_j), I((N' \cap N)', C_i, t_{j+1}), I\left((N' \cap N)', C_i, t_{j+j'}\right), \ldots, I((N' \cap N)', C_i, t_w)\right\} = 0, \forall \theta_{su} \tag{13}$$

This paper designs a network architecture which ensures that (13) holds true at all epochs.

The discussion so far assumes that data access from the cloud in the contexts considered above is accompanied with a high QoS. This assumes the availability of reliable network infrastructure. For instance, this assumption is not true where exists poor availability of high-performance terrestrial network infrastructure, or subscribers' inability to access expensive satellite networks. This assumption is true for cloud service providers in nations with a high population demanding access to cloud content; described by the conditions:

$$\left\{I(\beta = n_T^u, C_i, t_j), I(\beta = n_T^u, C_i, t_{j+1}), I\left(\beta = n_T^u, C_i, t_{j+j'}\right), \ldots, I(\beta = n_T^u, C_i, t_w)\right\} = 0, \forall n_T, \theta_g \tag{14}$$

$$\left\{I(\beta = n_S^r, C_i, t_j), I(\beta = n_S^r, C_i, t_{j+1}), I\left(\beta = n_S^r, C_i, t_{j+j'}\right), \ldots, I(\beta = n_T^u, C_i, t_w)\right\} = 0, \forall n_S \tag{15}$$

If we let $D(s_x, t_j)$ and $Th(s_x, \beta, \ t_j), \beta \epsilon\{\ n_T^u, n_S^r\}$ denote, respectively, the size of data accessed by s_x from network entity β and throughput associated with data accessed by s_x; $s_x \epsilon S$ via network entity β at epoch t_j, then the mean latency $\mathbb{E}(s_x)$ can be expressed as:

$$\mathbb{E}(s_x) = \frac{1}{2}\left(\frac{1}{wp}\sum_{j=1}^{w}\sum_{r=1}^{p}\frac{I(\beta = n_S^r, C_i, t_j) D(s_x, t_j)}{Th(s_x, \beta = n_S^r, t_j)} + \frac{1}{wq}\sum_{j=1}^{w}\sum_{u=1}^{q}\frac{I(\beta = n_T^q, C_i, t_j) D(s_x, t_j)}{Th(s_x, \beta = n_T^q, t_j)}\right) \tag{16}$$

Given the threshold latency l_{th}, the subscriber s_x has a significant delay if $\mathbb{E}(s_x) \gg l_{th}$. The delay $\mathbb{E}(s_x)$ refers to that of a single subscriber. In the case of multiple subscribers, the latency $\mathbb{E}_1(s_z)$ is given as:

$$\mathbb{E}_1(s_z) = \frac{1}{2}\left(\frac{1}{wpz}\sum_{j=1}^{w}\sum_{r=1}^{p}\sum_{z=1}^{x}\left(\frac{I(\beta = n_S^r, C_i, t_j) D(s_z, t_j)}{Th(s_x, n_s^r, t_j)}\right) + \ddot{\Upsilon}_1\right) \tag{17}$$

$$\ddot{\Upsilon}_1 = \frac{1}{wpz}\sum_{j=1}^{w}\sum_{u=1}^{q}\sum_{z=1}^{x}\left(\frac{I(\beta = n_S^r, C_i, t_j) D(s_z, t_j)}{Th(s_x, n_s^r, t_j)}\right) \tag{18}$$

There is a significant degradation associated with accessing cloud-based content when $\mathbb{E}_1(s_z) \gg l_{th}$,. Hence, a solution which ensures that the condition $\mathbb{E}_1(s_z) \leq l_{th}$ holds true for a significantly long duration is required. Such a solution is proposed in this paper. This section presents the two challenges being addressed in this paper, namely:

1. Ensuring that space tourist subscribers engaged in sub– orbital space tourism flight have continued access to internet and cloud-based content. For example, space tourist subscribers should be able to upload content observed at high altitudes and in outer space to the cloud with low latency.

2. Designing a solution which ensures that the condition $\mathbb{E}_1(s_z) \leq l_{th}$ holds true for subscribers desiring to access cloud-based content and for a significantly long duration.

3. Proposed solution(s) and associated mechanisms

This section presents the proposed solutions and is divided into two parts. The first part presents the solutions, mechanisms and associated network architecture to address the challenge involving space tourist subscribers. The second part discusses the solution that aims at ensuring the delivery of cloud-based content to subscribers at low latency when $\mathbb{E}_1(s_z) \gg l_{th}$.

3.1. Internet access continuity in space tourism

Space tourism subscribers are conveyed in a space vehicle that hosts communication subsystems which enables internet access. The space vehicle begins its journey from a terrestrial location with access to terrestrial wireless networks. The space tourists have access to the internet via the gateway of the terrestrial wireless network. In the context of the LTE-A utilizing the eNB, the P-GW and S-GW are the gateway entities. A handover is required to ensure the continuity of internet access as the space vehicle travels from the terrestrial location to outer space. Three handover levels are required in the proposed solution, these are:

1. **Terrestrial Wireless Network Handover (TWNH):** The TWNH refers to the handover executed between base station entities i.e. eNB. It is executed using protocols such as the seamless mobile internet protocol version 6 (SMIPv6) [41], mobile internet protocol version 6 (MIPv6) and proxy mobile internet protocol version 6 (PMIPv6) [42, 43]. The handover context implied in TWNH has been sufficiently addressed in literature.

2. **Terrestrial Wireless Network to Aerial Network Handover (TWAH):** The TWAH is necessary if the space vehicle connects to an aerial platform such as a high altitude platform as it sojourns to outer space. It involves the handover of a session from terrestrial wireless networks to aerial platforms. The aerial platforms in this context are connected using inter-platform links. Existing protocols such as that in [44] address the challenge of executing handover between terrestrial wireless networks and high altitude platforms.

3. **Aerial Platform to Satellite Handover (APSH):** The APSH involves executing a handover to the satellite on the uplink when subscribers access data from the cloud. The execution of

the APSH becomes necessary as the space vehicle's altitude increases as it approaches low earth orbit. Existing approach consider that satellites should handover to stratospheric platforms in reaching the subscribers. The case here is different because the satellite network is in the last mile.

4. **Sub-orbital Intersatellite Handover (SISH):** The execution of the SISH is required to ensure that the in-orbit space vehicle connects to the satellite enabling it to have the highest throughput and lowest latency. The SISH redefines the role of satellites in accessing cloud based information via the internet. This is because inter-satellite links have often been used with the aim of achieving global coverage using satellite networks; and not in the context of providing seamless high QoS internet connections to subscribers as a last mile technology.

The space tourist subscriber requires internet access for obtaining content from the internet or storing content for storage and later access. This should be realized without significant space segment acquisition costs. The contexts implied in the APSH and SISH require novel mechanisms and accompanying network architecture. This is because the APSH and SISH phases are peculiar to the space tourist subscriber. The relations between the TWNH, TWAH, APSH and SISH are shown in **Figure 2**.

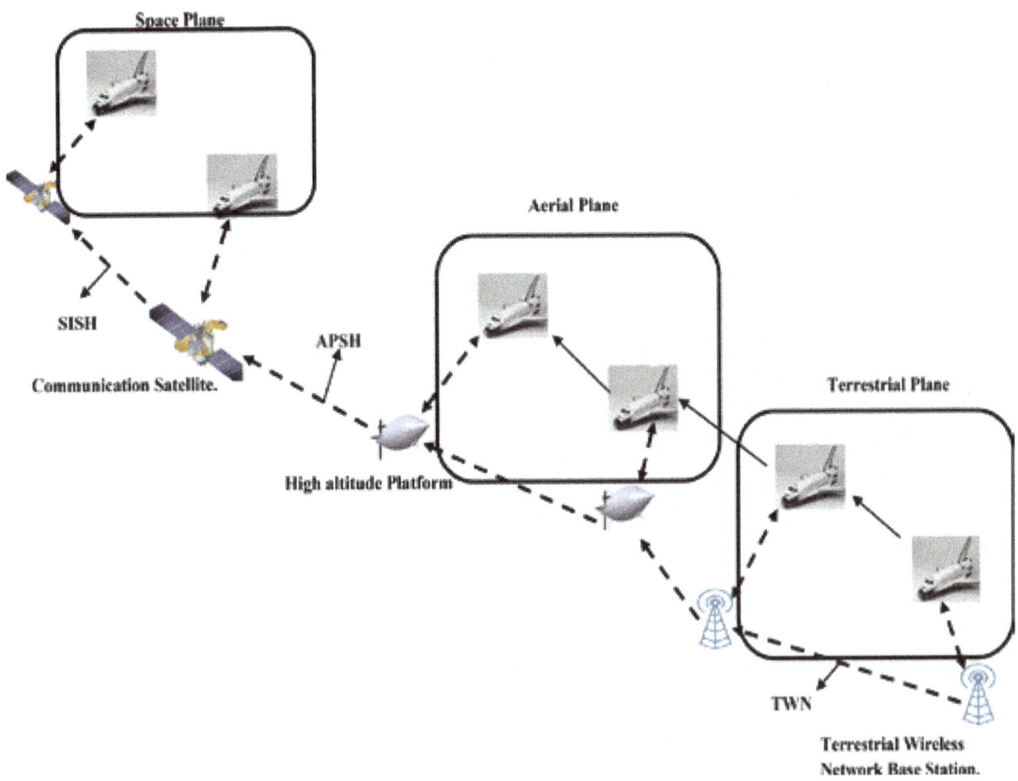

Figure 2. Relations between the TWNH, TWAH, APSH and SISH.

The scenario in **Figure 2** shows a space vehicle sojourning from a terrestrial location to outer space. The space vehicle passes through the terrestrial plane, aerial plane and the space plane. The space vehicle(s) is connected to the terrestrial wireless network base stations and access the cloud based content via the internet through the gateways. In the aerial plane, the space vehicle is connected to the high altitude platform.

The high altitude platform receives contents from select ground based stations. These ground stations are those being used for radio astronomy. However, they are not engaged in receiving radio astronomy signals during epoch of use by high altitude platform. The use of such ground stations is feasible considering the emergence of multi-mode ground stations that can be used for radio astronomy and packet processing [45]. The multi-mode ground station is connected to the computing infrastructure of the astronomy organization. The computing infrastructure is linked to the cloud computing platform hosting the content to be accessed by the space tourist subscriber. Idle multi-mode ground stations relay cloud based content to the space vehicle in the space plane. The cloud platform sends the cloud content to be sent to select ground stations that communicate with stratospheric platforms in the aerial plane. The select ground stations are also used to enable communications between satellites and the space vehicle in the space plane.

Aerial platforms communicate with each other using inter-platform links that utilize free space optics to ensure low latency. This is done when the space vehicle moves from the coverage of a high altitude platform to the coverage of another high altitude platform. In moving from the aerial plane to the space plane i.e. executing the APSH, the space vehicle does not have line of sight and communicates with the satellite via the ground station. In the space plane, the space vehicle is able to move between satellites. This is enabled by satellite communications with selected ground stations.

The proposed handover mechanism requires that the space vehicle conveying space tourists pass overhead through radio astronomy observatories. This provides the added benefit of enhancing astro-tourism and enables space tourists to have an aerial view of astronomical observatories. The re-use of existing astronomy infrastructure [46] reduces the cost associated with launching an anchor satellite to maintain high QoS internet connectivity for the concerned space vehicle. The use of selected ground station infrastructure improves the revenue potential for astronomy organizations; and increases the utilization of the high performance infrastructure and ground stations. The space vehicle connects to a geostationary communications satellite [47–51]. The ground segment of the geostationary satellite is an idle multi-mode ground station.

The handover algorithm that enables the provision of seamless internet connectivity for the space vehicle comprises entities that function in the space vehicle, ground stations, high altitude platforms and satellites. The proposed distributed handover algorithm (DHA) functions are for the aerial and space modes. The DHA executes the TWAH and the SISH in the aerial mode and space mode, respectively.

The space vehicle host mechanisms that enable it to execute the TWAH, and the handover between aerial platforms. However, these mechanisms are not designed since they have received considerable research attention. Let γ, Þ and $\mathfrak{p}$ be the set of terrestrial wireless network base station entities, stratospheric platforms and satellites, respectively.

$$\gamma = \{\gamma_1, \gamma_2, \ldots, \gamma_d\} \tag{19}$$

$$Þ = \{Þ_1, Þ_2, \ldots, Þ_h\} \tag{20}$$

$$\mathfrak{p} = \{\mathfrak{p}_1, \mathfrak{p}_2, \ldots, \mathfrak{p}_n\} \tag{21}$$

In addition, let $P(\varkappa, t_j), \varkappa \in \{\gamma_b, Þ_l, \mathfrak{p}_t\}, \gamma_b \in \gamma; Þ_l \in Þ; \mathfrak{p}_t \in \mathfrak{p}$ denote the strength of the signal form entity $\varkappa$ at epoch t_j. Given that $P_{th}(\gamma)$ is the threshold signal strength for terrestrial wireless network base station entity; the space vehicle measures the value of $P(\varkappa = \gamma_b, t_j)$ and $P(\varkappa = Þ_l, t_j)$ and retains connectivity to the terrestrial wireless network if:

$$\frac{1}{dw}\sum_{b=1}^{d}\sum_{j=1}^{w} P(\varkappa = \gamma_b, t_j) > P_{th}(\gamma) \tag{22}$$

$$\frac{1}{dw}\sum_{b=1}^{d}\sum_{j=1}^{w} P(\varkappa = \gamma_b, t_j) > \frac{1}{hw}\sum_{l=1}^{h}\sum_{j=1}^{w} P(\varkappa = Þ_l, t_j) \tag{23}$$

If (22) does not hold true, then (23) is also invalid. The APSH should be executed if:

$$\frac{1}{hw}\sum_{l=1}^{d}\sum_{j=1}^{w} P(\varkappa = Þ_l, t_j) > P_{th}(Þ) \tag{24}$$

$$\frac{1}{hw}\sum_{l=1}^{h}\sum_{j=1}^{w} P(\varkappa = Þ_l, t_j) > \frac{1}{hw}\sum_{b=1}^{d}\sum_{j=1}^{w} P(\varkappa = \gamma_b, t_j) \tag{25}$$

The space vehicle is in the terrestrial plane if (22), (23) hold true and is in the aerial plane when (24), (25) holds true. The space vehicle moves from the aerial to the space plane if:

$$\ddot{\Upsilon}_1 < \ddot{\Upsilon}_2 < \ddot{\Upsilon}_3 \tag{26}$$

$$\ddot{\Upsilon}_1 = \frac{1}{h \times (j + j')}\sum_{l=1}^{h}\sum_{j=1}^{j+j'} P(\varkappa = Þ_l, t_j) \tag{27}$$

$$\ddot{\Upsilon}_2 = \frac{1}{h \times (\alpha')}\sum_{l=1}^{h}\sum_{j=j+j'+1}^{j+j'+\alpha'} P(\varkappa = Þ_l, t_j) \tag{28}$$

$$\ddot{\Upsilon}_3 = \frac{1}{h \times (w - (j + j' + \alpha' + 1))}\sum_{l=1}^{h}\sum_{j=j+j'+\alpha'+1}^{w} P(\varkappa = Þ_l, t_j) \tag{29}$$

The transition in (26)–(29) involves a movement of the space vehicle from the aerial plane to the space plane. This handover is executed in the APSH. A set of relations describing the handover and the associated transition involving movement from the terrestrial plane to the aerial plane has not been presented. This kind of handover has been sufficiently addressed in the literature focused on aerial–terrestrial communications [45, 52, 53]. However, the context being addressed here is that of ensuring connectivity with a manned aerial vehicle (MAV) i.e. the space vehicle intended for space tourism.

The handover and transition implied in the SISH becomes activated when $P(\varkappa = Þ_l, t_j) < P(\varkappa = Þ_{l+1}, t_j); Þ_{l+1} \epsilon Þ$ and the space vehicle selects satellite $Þ_{l+1}$. The flowchart in **Figure 3** describes the relations executed in a handover procedure. The MAV searches for other networks of aerial platforms if the satellite signal is detected given that (26)–(29) holds true. In **Figure 3**, it is assumed that the space vehicle is able to connect to the concerned entities; i.e., high-altitude platforms or satellites depending on the decision context. The space vehicle connects to the entity with the highest transmit power.

3.2. Cyber: physical system - enhancing cloud access

The discussion presents a solution that enables subscribers to access cloud content when $\mathbb{E}_1(s_z) \leq l_{th}$. This scenario i.e. $\mathbb{E}_1(s_z) \leq l_{th}$ describes one in which terrestrial subscribers cannot access cloud content at low latency. In a terrestrial wireless network, a high latency arises when there is network congestion or network overloading. The occurrence of network congestion results in a low aggregate throughput in the network segment as well as a high latency. Existing research has considered the use of unmanned aerial vehicles to enhance the capacity of existing terrestrial wireless networks in several contexts [40, 41, 44]. Hence, unmanned aerial vehicles are suitable for addressing the challenge by providing an alternative path for accessing cloud content. Hence, unmanned aerial vehicles provide a cyber-physical extension (window) into the cloud platform. The proposed cyber-physical cloud comprises a central cloud platform or data center with several cloud extensions (windows). The use of the cyber-physical cloud also enhances the ability of space tourists to access cloud content. In this case, the space tourist is in the terrestrial plane when the condition $\mathbb{E}_1(s_z) \leq l_{th}$ is observed to hold true. Hence, the proposed cyber-physical cloud access system also enhances the provision of cloud based content to the internet.

In the cyber-physical cloud, the central cloud platform connects to the terrestrial wireless network and the cyber-physical windows as shown in **Figure 4**. In **Figure 4**, the cloud connects to either the aerial vehicle or the terrestrial wireless network. The subscribers desiring access to cloud based content are connected to the base station or the aerial vehicle. In the event that $\mathbb{E}_1(s_z) \gg l_{th}$, the notification is sent to the cloud and aerial vehicles are deployed.

The condition $\mathbb{E}_1(s_z) \gg l_{th}$ is verified at the cloud platform using information on the latency associated with data reception by each individual desiring access to cloud content. Each subscriber receiving content from the cloud via own terminals send information on the latency associated with content reception to the cloud platform. The usage of the term aerial vehicle implies both manned aerial vehicles (MAVs) and unmanned aerial vehicles (UAVs). The joint usage differs from the approach where only UAVs are used in the system [54].

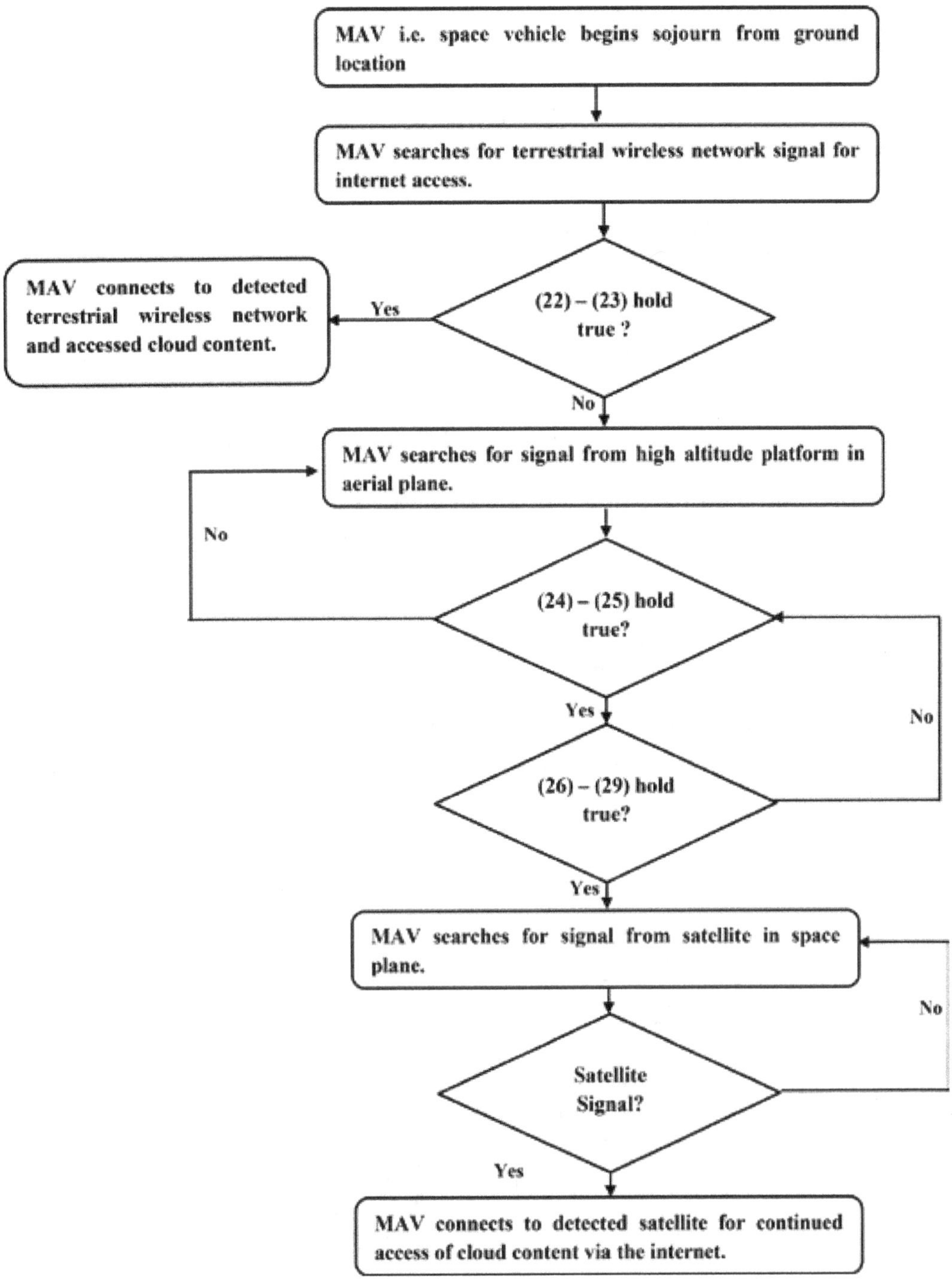

Figure 3. Functional flowchart showing the execution of the proposed handover for the space vehicle i.e. MAV.

The sole use of UAVs in the absence of aerial diversity does not consider regional aviation safety concerns. The incorporation of MAVs with UAVs enables the use of aerial vehicles in a manner that meets aviation safety concerns. For example MAVs are human driven and can be used in areas with constraints on aviation safety. The MAV is an aerial vehicle with smaller dimensions than the conventional manned aircraft; it is equipped with a communication payload that enables data communication with the cloud platform.

In the cyber–physical cloud, the central cloud platform connects to the terrestrial wireless network and the cyber–physical windows as shown in **Figure 4**. In **Figure 4**, the cloud can connect to either the aerial vehicle or the terrestrial wireless network. The subscribers desiring access to cloud-based content are connected to the base station or the aerial vehicle. In the event that $\mathbb{E}_1(s_z) > l_{th}$, the notification is sent to the cloud and aerial vehicles are deployed.

The condition $\mathbb{E}_1(s_z) > l_{th}$ is verified at the cloud platform using the information on the latency associated with data reception by each individual desiring access to cloud content. Each subscriber receiving content from the cloud via own terminals sends information on the latency associated with content reception to the cloud platform. The usage of the term aerial vehicle implies both MAVs and UAVs. The joint usage differs from the approach where only UAVs are used in the system [54]. The sole use of UAVs in the absence of aerial diversity does not consider stringent regional aviation safety concerns. The incorporation of MAVs with UAVs enables the use of aerial vehicles in a manner that meets stringent aviation safety concerns. For example, MAVs are human driven and can be used in areas with stringent constraints on aviation safety. The MAV is an aerial vehicle with smaller dimensions than conventional manned aircraft. It is equipped with a communication payload that enables data communication with the cloud platform.

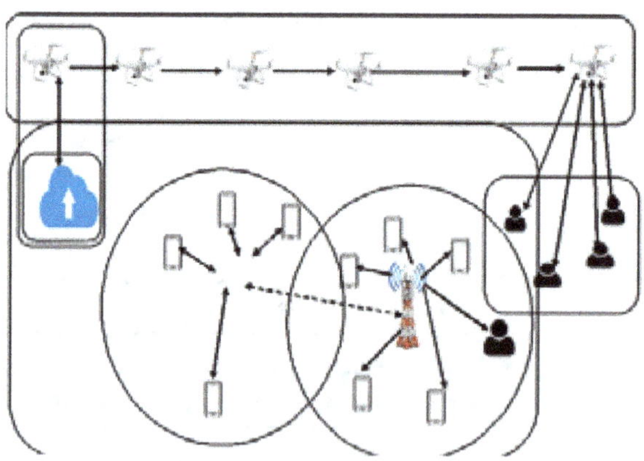

Figure 4. Incorporation of MAV and UAV into enabling cloud access.

4. Performance modeling and formulation

This section focuses on formulating the performance model of the proposed mechanisms. It is divided into two parts. The first part formulates the performance metrics of the mechanism enabling the internet access in space tourist applications. The second part formulates the performance analysis for the cyber-physical cloud system incorporating aerial diversity.

4.1. Performance model: pace tourist enabling mechanism

The formulated QoS metrics are the channel capacity and latency. The channel capacity for the space vehicle is formulated considering cases where the proposed handover mechanism is used and not used. The channel capacity achievable by the space vehicle in the terrestrial plane, aerial plane and space plane are denoted as C_{ter}, C_{ae} and C_{sp} respectively and can be expressed as:

$$C_{ter} = \sum_{z'=1}^{z'} \sum_{b=1}^{d} B(z', b) \, log_2 \left(1 + \frac{P_{tr}(\gamma_b, z')|h_{11}(\gamma_b, z')|^2}{P_{int}(\gamma_b, z')|h_{12}(\gamma_b, z')|^2 + \sigma^2} \right) \tag{30}$$

$$C_{ae} = \sum_{z'=1}^{z'} \sum_{l=1}^{h} B(z', l) \, log_2 \left(1 + \frac{P_{tr}(\text{Þ}_l, z')|h_{11}(\text{Þ}_l, z')|^2}{P_{int}(\text{Þ}_l, z')|h_{12}(\text{Þ}_l, z')|^2 + \sigma^2} \right) \tag{31}$$

$$C_{sp} = \sum_{z'=1}^{z'} \sum_{c=1}^{n} B(z', c) \, log_2 \left(1 + \frac{P_{tr}(\mathrm{p}_c, z')|h_{11}(\mathrm{p}_c, z')|^2}{P_{int}(\mathrm{p}_c, z')|h_{12}(\mathrm{p}_c, z')|^2 + \sigma^2} \right); \mathrm{p}_c \epsilon \, \mathrm{p} \tag{32}$$

Where:

z' is the channel z' which is distinct for each concerned communication entity.

$P_{tr}(\gamma_d, z')$, $P_{tr}(\text{Þ}_l, z')$ and $P_{tr}(\mathrm{p}_c, z')$ are the operational data transmit power between the space vehicle and (i) terrestrial wireless network on channel z', (ii) high altitude platform on channel z' and (iii) communication satellite on channel z' respectively.

$P_{int}(\gamma_d, z')$, $P_{int}(\text{Þ}_l, z')$ and $P_{int}(\mathrm{p}_c, z')$ is the interference power between the space vehicle and (i) terrestrial wireless network on channel z', (ii) high altitude platform on channel z' and (iii) communication satellite on channel z' respectively.

$h_{11}(\gamma_d, z')$, $h_{11}(\text{Þ}_l, z')$ and $h_{11}(\mathrm{p}_c, z')$ are the transmit channel gain between the space vehicle and (i) terrestrial wireless network on channel z', (ii) high altitude platform on channel z' and (iii) communication satellite on channel z' respectively.

$h_{12}(\gamma_d, z')$, $h_{12}(\text{Þ}_l, z')$ and $h_{12}(\mathrm{p}_c, z')$ are the transmit channel gain between the space vehicle and (i) terrestrial wireless network on channel z', (ii) high altitude platform on channel z' and (iii) communication satellite on channel z' respectively.

The average channel capacity of the space vehicle is denoted as C_{ave} and given as:

$$C_{ave} = \frac{1}{3}\left(C_{ter} + \ I(\gamma_d, Þ_l)C_{ae} + I(Þ_l, p_c)C_{sp}\right) \tag{33}$$

Where:

$I(\gamma_d, Þ_l)\epsilon\ \{0, 1\}$ is the handover indicator between γ_d and $Þ_l$. The cases $I(\gamma_d, Þ_l) = 0$ and $I(\gamma_d, Þ_l) = 1$ signify that a handover is not executed and is executed between γ_d and $Þ_l$ respectively.

$I(Þ_l, p_c)\epsilon\ \{0, 1\}$ is the handover indicator between $Þ_l$ and p_c. The cases $I(Þ_l, p_c) = 0$ and $I(Þ_l, p_c) = 1$ signify that a handover is not executed and is executed between $Þ_l$ and $p_c$respectively.

The latency associated with transmitting D' bytes of data without and with the incorporation of the proposed handover is denoted β_1 and β_2 respectively and given as:

$$\beta_1 = 8 \times D' \times (C_{ave})^{-1}\Big|_{I(\gamma_d, Þ_l)=0,\ \ I(Þ_l, p_c)=0} \tag{34}$$

$$\beta_2 = 8 \times D' \times (C_{ave})^{-1}\Big|_{I(\gamma_d, Þ_l)=1,\ \ I(Þ_l, p_c)=1} \tag{35}$$

The cases $8 \times D' \times (C_{ave})^{-1}\Big|_{I(\gamma_d, Þ_l)=1,\ \ I(Þ_l, p_c)=0}$ and $8 \times D' \times (C_{ave})^{-1}\Big|_{I(\gamma_d, Þ_l)=0,\ \ I(Þ_l, p_c)=1}$ have not been considered. This is because our discussion does not consider a partial handover as implied in the cases described by $I(\gamma_d, Þ_l) = 1,\ \ I(Þ_l, p_c) = 0$ and $I(\gamma_d, Þ_l) = 1,\ \ I(Þ_l, p_c) = 1$. A partial handover results in a scenario where QoS of the space tourist subscribers suffer severe degradation due to frequent interruption.

4.2. Performance model: cyber-physical aided cloud access system

The deployment of either the UAV or MAV in the proposed cyber–physical cloud access system enables the delivery of cloud content when it could otherwise be challenging. This is due to the incidence of network congestion or any other event that could lead to high delay in the terrestrial wireless network segment. In the formulation, the cloud content traverses multiple cells in an infrastructure-based network. The occurrence of congestion on any of the forwarding network nodes increase the latency associated with accessing cloud content by remote subscribers. The probability of congestion on terrestrial wireless network γ_b with own base station and associated gateway entity is denoted $P_{co}(\gamma_b); \gamma_b\ \epsilon\ \gamma$. The probability of deploying either MAVs or UAVs that spans the coverage of $|\gamma|$ terrestrial wireless networks is denoted as $P_{cy}(|\gamma|)$. Given that the channel capacity of γ_b is denoted as $Th(\gamma_b)$; the aggregate channel capacity associated with cloud content without and with the cyber–physical system is denoted as $Th_1^{cl}(\gamma_b)$ and $Th_2^{cl}(\gamma_b)$, respectively.

$$Th_1^{cl}(\gamma_b) = \left(\prod_{b=1}^{f} P_{co}(\gamma_b) \times \sum_{b=f+1}^{d} P_{co}(\gamma_b)\right) \times Th(\gamma_b)\forall\gamma_b \tag{36}$$

$$Th_2^{cl}(\gamma_b) = \left(\left(\prod_{b=1}^{f} P_{co}(\gamma_b) \times \sum_{b=f+1}^{d} P_{co}(\gamma_b)\right) + \left(\prod_{b=1}^{f} P_{cy}(\gamma_b) \times \sum_{b=f+1}^{d} P_{cy}(\gamma_b)\right)\right) \times Th(\gamma_b)\forall\gamma_b \tag{37}$$

5. Simulation and discussion of results

This section presents and discusses the simulation results and performance benefits of the proposed mechanisms. It is divided into three parts. The first part presents the simulation parameters for the proposed mechanisms. The second part presents results indicating the performance of the space tourist subscriber. The third part presents results on the proposed cyber-physical aided cloud system.

5.1. Simulation parameters

The simulation parameters used to investigate the performance benefit of the handover mechanism for the space tourist subscriber are shown in **Table 3**. The parameters used to investigate the performance of the cyber – physical cloud access system are shown in **Table 4**.

5.2. Discussion of results – space tourist

The results of performance simulation are presented in this subsection. The performance benefit of the proposed handover mechanism is investigated using the channel capacity and latency as metrics. In addition, the performance benefit of incorporating aerial diversity is investigated using the aggregate channel capacity. The proposed aerial diversity mechanism is used to improve access to cloud content in terrestrial wireless networks.

S/N	Parameter	Value
1	Mean of transmit power $P_{tr}(\gamma_d, z')$ for the space vehicle.	202.9 mW
2	Mean of interference power for space vehicle in terrestrial plane, $P_{int}(\gamma_d, z')$	10.3 mW
3	Channel bandwidth in terrestrial plane, $B(z', b)$	1.5 MHz
4	Number of channels that are simultaneously accessed in terrestrial plane	4
5	Mean of transmit power of space vehicle in aerial plane, $P_{tr}(Ꝑ_l, z')$	313 mW
6	Mean of interference power of space vehicle in aerial plane, $P_{int}(Ꝑ_l, z')$	38.9 mW
7	Channel bandwidth in aerial plane	2.25 MHz
8	Number of channels that are simultaneously accessed in aerial plane	4
9	Mean of space vehicle transmit power $P_{tr}(p_c, z')$ in space plane.	227.6 mW
10	Average space vehicle interferer power in space plane, $P_{int}(Ꝑ_c, z')$	32.6 mW
11	Channel bandwidth in space plane	5 MHz
12	Number of channels that are simultaneously accessed in space plane	4

Table 3. Parameters used to investigate the performance of the handover mechanism.

	Epoch 1			Epoch 2		
Base station index	**Mean congestion probability**	**Mean MAV, UAV deployment probability**	**Mean backhaul throughput (Gbps)**	**Mean congestion probability**	**Mean MAV, UAV deployment probability**	**Mean backhaul throughput (Gbps)**
1	0.3888	0.5685	35.3279	0.3664	0.4411	29.3312
2	0.7627	0.6175	29.5502	0.5061	0.4600	34.6044
3	0.5962	0.4267	19.5083	0.5352	0.5510	29.7344
4	0.7723	0.5495	21.4568	0.7274	0.5709	42.5083
5	0.6186	0.5712	22.3693	0.5334	0.5342	28.5267
6	0.6430	0.4176	35.3991	0.5230	0.6319	22.5061
7	0.3911	0.5927	30.0374	0.3145	0.6141	34.6840
8	0.3109	0.4928	31.4955	0.4969	0.5486	27.6927
9	0.4991	0.4411	21.9782	0.2291	0.4413	25.4814
10	0.5953	0.4336	49.2761	0.2747	0.4072	41.4290
11	0.4421	0.5791	28.4492	0.5575	0.6513	33.4003
12	0.6049	0.4986	41.8964	0.1730	0.5996	41.9393
13	0.3588	0.4654	13.7261	0.1503	0.2617	12.4052
14	0.5273	0.3811	28.5571	0.4534	0.3425	25.8474
15	0.5528	0.6437	45.3798	0.5048	0.5061	38.7138
16	0.6340	0.3282	11.0572	0.3233	0.5900	8.7650
17	0.3755	0.4869	25.0408	0.3963	0.5300	29.4266
18	0.5843	0.4820	34.9777	0.6028	0.4556	29.6869
19	0.3843	0.5792	30.5558	0.5304	0.5177	37.5723
20	0.5794	0.4632	29.2089	0.5655	0.5562	35.5610
21	0.6785	0.4600	31.2913	0.5464	0.5881	25.5595
22	0.2695	0.4369	37.4373	0.6859	0.6047	38.6078
23	0.5935	0.4550	40.7390	0.5582	0.4034	29.8772
24	0.3468	0.5875	35.4107	0.4412	0.4862	38.5071
25	0.4491	0.4572	39.0279	0.3655	0.4827	33.7495
26	0.4212	0.5881	27.4572	0.2794	0.5679	34.3914
27	0.4929	0.5598	30.8914	0.5723	0.5080	25.5206
28	0.5301	0.3996	34.4666	0.5208	0.6065	33.9866
29	0.5138	0.5482	21.8294	0.3015	0.5177	29.1524
30	0.4620	0.5018	42.4102	0.4737	0.4948	51.8990
31	0.5426	0.5050	18.7730	0.6504	0.3931	15.3598
32	0.6005	0.4215	35.4899	0.5052	0.5537	28.0000
33	0.1818	0.6061	38.9559	0.1964	0.5787	39.6352
34	0.5360	0.6202	10.5887	0.5137	0.6282	6.8601

	Epoch 1			Epoch 2		
Base station index	**Mean congestion probability**	**Mean MAV, UAV deployment probability**	**Mean backhaul throughput (Gbps)**	**Mean congestion probability**	**Mean MAV, UAV deployment probability**	**Mean backhaul throughput (Gbps)**
35	0.5848	0.4162	35.9282	0.4419	0.5076	29/9443
36	0.2103	0.5099	22.9408	0.4989	0.6247	38.3689
37	0.2227	0.6742	21.7225	0.5790	0.5600	44.0250
38	0.7069	0.6086	37.3177	0.6596	0.4421	27.1582
39	0.4296	0.6832	39.0351	0.4496	0.6709	34.6300
40	0.4539	0.4368	34.1860	0.4793	0.6442	36.4133

Table 4. Simulation parameters – cyber – physical aided cloud access mechanism.

Simulation result for the space vehicle average channel capacity is presented in **Figure 5**. **Figure 5** shows two sub-figures, i.e. a and b. The average channel capacity before and after the incorporation of the proposed handover mechanism is presented in **Figure 5a** and **b**,

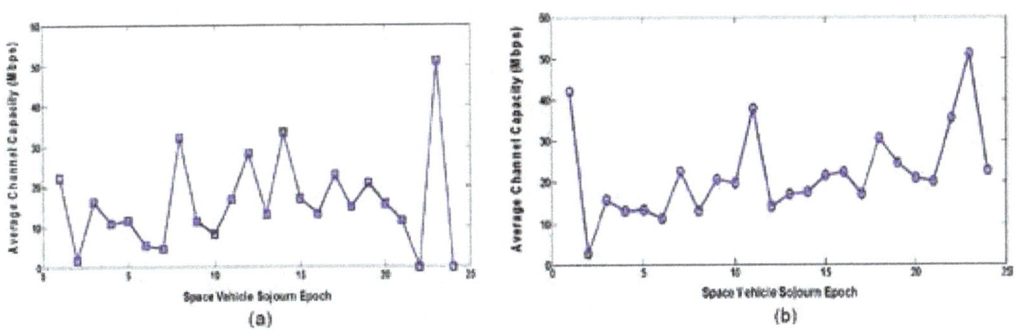

Figure 5. Average channel capacity of the space vehicle before and after introducing the proposed scheme. (a) Average channel capacity achieved by space vehicle in Mbps in the absence of the proposed handover mechanism. (b) Average channel capacity achieved by space vehicle in Mbps after introducing the proposed handover mechanism.

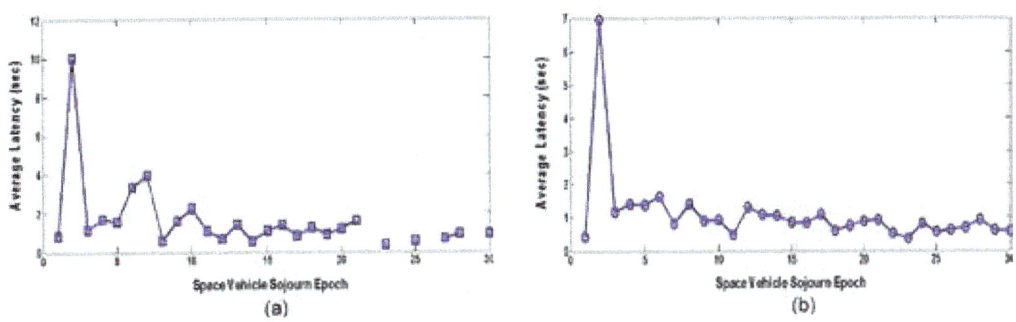

Figure 6. Average latency of the space vehicle before and after introducing the proposed scheme. (a) Average latency achieved by space vehicle without the proposed handover mechanism. (b) Average latency achieved by space vehicle after the incorporation of the proposed handover mechanism.

respectively. Analysis of the result shows that the incorporation of the proposed mechanism enhances channel capacity. This is because of the continuity in data transmission during space vehicle sojourn. It is observed from the results that the channel capacity is enhanced on average by 18.4%.

The result for the latency of the space vehicle is presented in **Figure 6a** and **b**. **Figure 6a** and **b** shows the latency without and with the proposed mechanism, respectively. Results show that the proposed mechanism reduces the latency associated with accessing cloud content by the space vehicle. Analysis shows that the proposed handover mechanism reduces latency on average by 12%.

Aggregate channel capacity (bits per second)				
Epoch 1			**Epoch 2**	
Number of base stations	**Without aerial diversity**	**With aerial diversity**	**Without aerial diversity**	**With aerial diversity**
2	6.76×10^{4}	3.85×10^{4}	1.88×10^{5}	3.94×10^{5}
4	1.61×10^{11}	1.56×10^{12}	8.78×10^{12}	1.72×10^{13}
6	1.73×10^{19}	4.10×10^{19}	1.73×10^{20}	3.54×10^{20}
8	3.33×10^{26}	7.87×10^{26}	1.89×10^{27}	5.75×10^{27}
10	4.45×10^{33}	1.23×10^{34}	1.10×10^{34}	5.44×10^{34}
12	1.01×10^{41}	1.92×10^{41}	2.05×10^{41}	1.15×10^{42}
14	5.74×10^{47}	1.10×10^{48}	3.82×10^{47}	3.45×10^{48}
16	6.40×10^{54}	1.66×10^{55}	7.47×10^{54}	6.75×10^{55}
18	9.49×10^{61}	2.62×10^{62}	8.71×10^{61}	1.02×10^{63}
20	1.18×10^{69}	3.34×10^{69}	1.74×10^{69}	1.99×10^{70}
22	1.16×10^{76}	3.02×10^{76}	2.42×10^{76}	2.98×10^{77}
24	1.94×10^{83}	4.54×10^{83}	4.04×10^{83}	3.71×10^{84}
26	3.17×10^{90}	6.85×10^{90}	8.46×10^{90}	7.67×10^{91}
28	9.40×10^{89}	1.53×10^{90}	1.26×10^{90}	1.42×10^{91}
30	9.40×10^{89}	1.18×10^{97}	1.26×10^{90}	2.07×10^{98}
32	7.61×10^{96}	9.06×10^{103}	1.11×10^{97}	1.29×10^{105}
34	5.85×10^{103}	9.87×10^{110}	1.11×10^{104}	2.80×10^{112}
36	5.74×10^{110}	2.17×10^{125}	8.62×10^{110}	6.46×10^{126}
38	1.31×10^{125}	5.59×10^{132}	1.74×10^{125}	1.59×10^{134}
40	3.69×10^{132}	7.34×10^{139}	4.43×10^{132}	3.67×10^{141}
Mean improvement in aggregate channel capacity		**70.3634%**	**Mean improvement in aggregate channel capacity**	**85.1583%**

Table 5. Aggregate Channel capacity before and after incorporating aerial diversity.

The investigation also examines how aerial diversity enhances cloud content access. The use of aerial diversity ensures that network congestion does not affect the ability of the space vehicle subscriber to access cloud content. The metric used to investigate the performance benefit of incorporating aerial diversity is the aggregate channel capacity. This is the achieved channel capacity when the access of cloud content requires communications between multiple base stations. This is investigated using the parameters in **Table 4**. The aggregate channel throughput obtained via simulations for two epochs (i.e. epoch 1 and epoch 2) is shown in **Table 5**.

The simulation results in **Table 5** show that incorporating aerial diversity enhances the aggregate channel capacity. The incorporation of aerial diversity enables the delivery of cloud content when the terrestrial wireless network experiences congestion. The aggregate channel capacity is increased as aerial diversity influences data transmission for an increasing number of base stations. This is because the MAV and UAV are deployed in a manner that enables the delivery of cloud content for a larger terrestrial wireless network coverage area. Therefore, aerial diversity enhances the aggregate channel capacity associated with accessing cloud content by the space vehicle subscribers as seen in epochs 1 and 2. Aerial diversity enhances the aggregate channel capacity by 70.4 and 85.2% on average at epoch 1 and epoch 2, respectively.

6. Conclusion

This paper has proposed a distributed handover algorithm (DHA) for a network comprising terrestrial, aerial and space-based segments. DHA enables aerial platform to satellite handover and suborbital intersatellite handover in a space vehicle. The network uses ground stations deployed for radio astronomy. Simulation shows that DHA enhances channel capacity and reduces latency by 18.4 and 11.6% on average, respectively. The paper also proposed the joint use of manned and unmanned aerial vehicles for improving accessibility to cloud content while minimizing aviation safety concerns. The aerial vehicles are deployed when cloud access via terrestrial wireless networks is subject to significant latency due to network congestion. Simulations show that using aerial diversity reduces the effect of network congestion on terrestrial wireless networks. The aggregate throughput achieved on the terrestrial wireless network increases by up to 85% on average.

Author details

Ayodele Abiola Periola

Address all correspondence to: ayperiola@bellsuniversity.edu.ng

Department of Electrical, Electronics and Computer Engineering, Bells University of Technology, Ogun State, Nigeria

References

[1] Zhang Y, Zheng T, Dong P, Luo H, Pang Z. Comprehensive analysis on heterogeneous wireless networks in high speed scenarios. Wireless Communications and Mobile Computing. 2018;**2018**:4259510. pp. 1-12

[2] Tian Y, Xu K, Ansari N. TCP in wireless environments: Problems and solutions. IEEE Communications Magazine. 2005;**3**(3):S27-S32

[3] Porta TF, Satgarelli L, Forster GT. Mobile IP and wide area wireless data. In: IEEE Wireless Communications and Networking Conference. New Orleans, USA; 1999. DOI: 10.1109/WCNC/1999/796944

[4] Cisco. Cisco Virtual Networking Index: Global Mobile Data Traffic Forecast Update, 2016–2021, Document ID: 1454457600805266; 2017. pp. 1-35

[5] Ericsson. Ericsson Mobility Report; 2018. pp. 1-35

[6] Yadav KA, Kumar S. A review of congestion control mechanisms for wireless networks. In: International Conference on Communication and Electronic Systems. Coimbatore, India; 2017. DOI: 10.1109.CESYS.2017.8321245

[7] Kong Y, Zang H, Ma X. Improving TCP congestion control with machine intelligence. In: Proceedings of the 2018 Workshop on Network Meets AI and ML. Budapest, Hungart; 2018. pp. 60-66

[8] Winstein K, Balakrishnan H. TCP ex machina: Computer generated congestion control. In: Proceedings of the ACM SIGCOMM, 2013. Hong Kong; 2013. pp. 123-124

[9] Papadimitriou D, Welzl M, Scharf M, Briscoe B. Open research issues in internet congestion control. Internet Research Task Force–Request for Comments – 6077. 2011. pp. 1-51

[10] Poojary S, Sharma V. Analysis of Multiple Flows using different high speed TCP protocols on a general network. Available from: https://arxiv.org/pdf/1602.06652.pdf [Accessed: 26 September 2018]

[11] Denning PJ. The ARPANET after twenty years. American Scientist. 1989;**77**(6)

[12] Atkinson RD, Bennett R. The Future of the Internet and Broadband and How to Enable It. Available from: www.itif.org/files/120090903_the_%_20_future_of_the_interner_file.pdf [Accessed: 30 September 2018]

[13] Bay M. Conversations with a pioneer: Paul Baran in his own words. Internet Histories. 2017;**1**(3):273-282. DOI: 10.1080/24701475.2017.1345469

[14] Lee TB. Information Management: A Proposal. 1989. Available from: https://cds.cern.ch/record_369245/files/ARCH_www_4_010.pdf [Accessed: 30 September 2018]

[15] Sullivan JO. The Role of Astronomy in Wi – Fi. Available from: www.atnf.csiro.au/research/conferences/2016/IDRA16/presentation/O_SullivanJohn.pdf [Accessed: 30 September 2018]

[16] Liu F. Definition and Research of Internet Neurology. Available from: https://arxiv.org/ftp/arxiv/paper1504/1504.-2842.pdf

[17] Liu F, Shi Y, Li P. Analysis of the relation between artificial intelligence and the internet from the perspective of brain science. Procedia Computer Science. 2017;**122**:377-383

[18] Sweeting MN. Modern small satellites – Changing the economics of space. Proceedings of the IEEE. 2018;**106**(3):343-361

[19] Ilcev SD. Stratospheric communication platforms as an alternative for space program. Aircraft Engineering and Aerospace Technology. 2011;**83**(2):105-111

[20] Vergouw B, Nagel H, Bondt G, Custers B. Drone technology: Types payloads, applications, frequency, spectrum issues and future developments. In: The Future of Drone Use. The Hague: T.M.C. Asser Press; 2016. pp. 21-45

[21] Bond P. Directorate of Human Spaceflight. Benefits of Human Spaceflight. Available from: www.esa.int/esapub/br/br231/br230.pdf

[22] STCI. Space Tourism Industry Forecast; 2018

[23] Cole S. Space tourism: Prospects, positioning and planning. Journal of Tourism Futures. 2015;**1**(2):131-140

[24] Toivonen A. Sustainable planning for space tourism. Matkailututkimus. 2017;**13**(1–2):21-34

[25] Mindell DA, Uebelhart SA, Gerovitch S, Hoffman J, Lanford E, Logsdon J, et al. The Future of Human Spaceflight. Available from: https://web.mit.edu.mitsps/MITFutureof-HumanSpaceflight.pdf [Accessed: 30 September 2018]

[26] Crawford IA. The scientific case for a human spaceflight infrastructure. Earth, Moon and Planets. 1999;**67**(3):221-231

[27] Available from: www.solstar.com/about/

[28] Available from: https://wifinowevents.com/news-and-blog/outer-space-gets-its-first-isp/

[29] Available from: https://www.virgingalactic.com

[30] Available from: https://www.mars-one.com/faq/technologies/how/does/themars-one/-communicate-with-earth

[31] Do S, Owens A, Ho K, Schreiner S, Week O. An independent assessment of the technical feasibility of the mars one mission plane – Updated analysis. Acta Astronautica. 2016;**120**:192-228

[32] Turton R. Multiplanetary imageries and utopia: The case of mars one. Science, Technology & Human Values. 2017. DOI: 10.1177/0162243917737366

[33] Ahmed T, Dubois E, Dupe J, Ferrus R, Gelard P, Kuhn N. Software defined satellite cloud RAN. International Journal of Satellite Communications and Networking. 2018;**36**:108-133

[34] Yang G, Lin X, Li Y, Cui H, Xu M, Wu D, et al. A Telecom Perspective on the internet of Drones from LTE – Advanced to 5G. Available from: https://arxiv.org/abs/1803.11048

[35] Shi W, Zhou H, Li J, Xu W, Zhang N, Shen X. Drone assisted vehicular networks: Architecture, challenges and opportunities. IEEE Network. 2018;**32**(3):130-137

[36] Sharma V, You I, Kul G. Socializing drones for inter – Service operability in Ultradense wireless networks using block chain. In: Proceedings of the 2017 International Workshop on Managing Insider Security Threats. Dallas, TX, USA; 2017. pp. 81-84

[37] Saldama J, Morat AA, Sathiaseelan A, Braem B, Pietrosemoli E, Zennaro M, et al. Alternative networks: Towards global access to the internet for all. IEEE Communications Magazine. 2017;**55**(9):187-193

[38] Rushing S. Analyzing Global Access to the Internet for All Projects. Seminar Future Internet, WS 2015/6, Network, Architectures and Services; 2016. pp 9-16

[39] Ahmadi H, Fontanesi G, Katzis K, Shakir MZ, Zhu A. Resilience of Airborne networks. Available from: https://arxiv.org/pdf/1807.10962 [Accessed: 27 September 2018]

[40] Purohith DR, Sivatangam KM. SIFM: Network architecture for seamless flow mobility between LTE and WiFi Networks - Analysis and Testbed implementation. Available from: https://arxiv.org/ftp/arxiv/papers/1702.1702.07489.pdf [Accessed: 27 September 2018]

[41] Yen YS, Chen LY, Chi TY, Chao HC. A novel predictive scheduling handover on mobile IPv6. Telecommunication Systems. February 2013;**52**:461-473

[42] Schmidt T, Waehlisch M, Koodli R, Fairhurst G, Liu D. Multicast listener extensions for mobile IPv6 and proxy mobile IPv6 (PMIPv6) fast handovers. In: Internet Engineering Task Force (IETF): Request for Comments 7411. 2014. pp. 1-30

[43] Perkins C, Devarapali V. Mobile node identifier types for MIPv6. In: Internet Engineering Task Force, Request for Comments 8371. 2018. pp. 1-16

[44] Chandrasekharan S, Gomez K, Hourani A, Kandepaan S, Rasheed T, Goratti L, et al. Designing and Implementing Future Aerial Communication Networks. Available from: https://arxiv.org/ftp/arxiv/papers/1602/1602.05318.pdf [Accessed: 28 September 2018]

[45] Periola AA, Falowo OE. Intelligent cognitive radio models for enhancing future radio astronomy observations. Advances in Astronomy. 2016:5408403. pp. 1-16

[46] Periola AA, Falowo OE. Instrumentation location diversity paradigm for future astronomy observations. Wireless Personal Communications. 2018:1-25. DOI: 10.1007/S 11277-018-5940-x

[47] Namuduri K. When Disaster Strikes, Flying Cell Towers could Aid Search and Rescue. IEEE Spectrum. 2017. Available from: https://spectrum.ieee.org.telecom/wireless/when-disaster-strikes-flying-cell-towers-could-aid-search-and-rescue [Accessed: 28 September 2018]

[48] Kitindi EJ, Fu S, Jia Y, Kabir A, Wang Y. Wireless network virtualization with SDN and C – RAN for 5G networks: Requirements, opportunities and challenges. IEEE Access. 2017;**5**:19099-19115

[49] Hollier S. The Accessibility of Cloud Computing – Current and Future Trends. Sydney: Media Access Australia; 2014. pp. 1-38

[50] Mujinga M, Chipangura B. Cloud computing concerns in developing economies. In: Australian Information Security Management Conference. Australia; 2011. pp. 196-203

[51] Hole KJ. In: Tveito A, editor. Anti – fragile ICT Systems. AG Switzerland: Springer International Publishing; 2016

[52] Shi Y, Liu J, Fadullah ZM, Kato N. Cross – Layer data delivery in satellite – Aerial – Terrestrial communications. IEEE Wireless Communications. June 2018;**25**(3):138-143

[53] Kandepaan S, Gomez K, Reynaurd L, Rasheed T. Aerial – Terrestrial communications terrestrial cooperation and energy efficient transmission to Aerial Base stations. IEEE Transactions on Aerospace and Electronic Systems. 2014;**50**(4):2715-2735

[54] Mahmoud S, Mohamed N, Jaroodi J. Integrating UAVs into the clouds using the concept of the web of things. Journal of Robotics. 2015;**2015**:631240. pp. 1-10

9

Autonomous Underwater Vehicle Guidance, Navigation and Control

Timothy Sands and Kevin Bollino

Abstract

A considerable volume of research has recently blossomed in the literature on autonomous underwater vehicles accepting recent developments in mathematical modeling and system identification; pitch control; information filtering and active sensing, including inductive sensors of ELF emissions and also optical sensor arrays for position, velocity, and orientation detection; grid navigation algorithms; and dynamic obstacle avoidance among others. In light of these modern developments, this article develops and compares integrative guidance, navigation, and control methodologies for the Naval Postgraduate School's Phoenix, a submerged autonomous vehicle. The measure of merit reveals how well each of several methodologies cope with known and unknown disturbance currents that can be constant or harmonic while maintaining safe passage distance from underwater obstacles, in this case submerged mines.

Keywords: submersible vehicles, ocean research, obstacle avoidance, guidance, navigation, and control, linear quadratic optimal control, approximated optimal control, reduced-order observers, MIMO, SIMO

1. Introduction

The Naval Postgraduate School's consortium for robotics and unmanned systems education and research (CRUSER) uses three autonomous underwater vehicles, the Remus, Aries [1], and Phoenix [2] vehicles to enhance education and research. The oldest vehicle, Phoenix [3] is used in this study to investigate integrated methodologies [4] for vehicle guidance, navigation, and control through a field of obstacles amidst unknown ocean currents that can be approximated by steady state, fixed disturbance ocean velocities, and can also be represented by harmonically oscillating velocities. This integrated approach is a natural extension of the recent innovations. The Phoenix vehicle's nominal mathematical modeling was articulated in the 1988 article [5] using surge motion to perform system identification. Recent innovations [6–10] have extended and improved the nominal system identification resulting in high-confidence mathematically modeling in computer simulations. Such simulations permitted Wu et al. [11] to redesign the L1 adaptive control architecture for pitch-control with anti-windup compensation based on solutions to the Riccati equation to guarantee robust and fast adaption of the underwater vehicle with input saturation and coupling disturbances and the approach was applied to the pitch channel alone. Stability was emphasized in the single-channel approach to emphasize dynamic nonlinearities and measurement errors. The Riccati equation is also utilized in this research and proves effective when applied to all six degrees of freedom per [4], where the approach is applied to instances of disturbances that are constant with simultaneous harmonic disturbances simulating unknown ocean currents and waves. In addition to these recent achievements in control, improvements have also been made to guidance and navigation. In recent years, Bo He et al. [12] demonstrated in simulations and open water experiments, the ability to overcome weak data links and sparse navigation data using a technique called extended information filter (EIF) applied to simultaneous localization and mapping (i.e. "SLAM") that proved computationally easier to implement than traditional extended Kalman filter (EKF) SLAM. Low computational cost is emphasized here to keep the vehicle size low, but also to exaggerate the laudable goal of achieving optimal or near optimal results with methods that are simple. Such is an overt goal of the new research presented here.

Just last year, Yan et al. [13] integrated the navigation system using a modified fuzzy adaptive Kalman filter (MFAKF) to combine traditional strap-down inertial navigation with OCTANS and Doppler velocity log (DVL) to navigate the challenging polar regions where rapidly converging earth meridians and challenging ocean environments filled with submersed obstacles. This benchmark achievement requires the research here to utilize similar challenging ocean conditions, and provide the motivation for selection of simultaneous steady-state ocean currents together with sinusoidal varying unknown wave conditions amidst an ocean filled with obstacles (where here the non-polar ocean is used, so mines are added to fulfill the role of malignant submersed obstacles). Furthermore, simplified waypoint guidance is derived, based on the onboard-calculated distance from the vehicle to a submerged obstacle. The simplified waypoint guidance is proven effective, and should be considered in situations where onboard operation of a modified fuzzy adaptive Kalman filter proves to be computationally prohibitive. The distance to an underwater obstacle was measured by Wang et al. [14] with a novel method: measuring extremely low frequency (ELF) emissions with onboard inductive sensors. Such emissions are produced by ship hulls with relatively pronounced amplitudes compared to small subsurface obstacles, but the harmonic line spectra and fundamental signal frequency relate directly to the closing speed of approach to the obstacle. Experiments proved that even such small signals were detectable at long range with high sensitivity and low-noise sensors of the current state of the art, thus closing distance to obstacles may now be presumed to be known passively, permitting the simplified waypoint guidance proposed in this manuscript. Particularly after ELF queuing, position, orientation, and velocity of obstacles may be monitored optically as developed by Eren et al. [15], and these states may be used as feedback signals together with the waypoint guidance (desired trajectory) permitting augmentation with linear quadratic Gaussian techniques, as done in this manuscript where full-order state observers are together optimized with attitude controller gains, followed by demonstration that reduced-order observers may also be optimized allowing vehicle operators to compensate for individually failed or degraded sensors, or instances where optimally estimated signals are superior to sensor signals in an individual or multiple channel.

Integrating these latest technological developments was demonstrated last year by Wei et al. [16], who integrated the Doppler velocity methods for obstacle monitoring into a dynamic obstacle avoidance scheme for collision avoidance. Following data fusion, a collision risk assessment model is used to avoid collisions, and claims to be effective in unknown dynamic environments, although the experiments did not go so far as to stipulate near-constant ocean currents in addition to harmonic wave actions. These challenging dynamic environments are addressed in this manuscript as a natural extension of the current state of the art.

Autonomous vehicle angular momentum control of rotational mechanics may be achieved using control moment gyroscopes, one potential momentum exchange actuator with a long, historic legacy actuating space vehicles, where mathematical singularities have just recently been overcome [17–23], permitting use of the actuator for underwater vehicles as done recently achieved by Thorton et al. [24, 25] including combined attitude and energy storage control. These developments suffice to reveal that attitude control is not controversial, and thus the remainder of this manuscript focuses on guidance and navigation with a residual necessity to implement nominal, effective pitch and yaw control.

2. Materials and methods

Submersible vehicles require control systems to guide the vehicle around obstacles that can present dangers to vehicle health and safety in the presence of ocean currents. The challenge addressed here is to navigate the Naval Postgraduate School's *Phoenix* submersible vehicle (**Figure 1**) through a minefield whose dimensions are 200 m × 5100 m in the presence of 0.5 m/s ocean currents. The field will contain at least 30 mines placed at locations using a random number generator. The resulting controller structure has an inner-outer loop structure, and several technologies will be described including pole-placement designs, linear-optimal (quadratic) Gaussian techniques, full and partial order observers for online disturbance identification for ocean currents (both constant lateral underwater ocean currents and also sinusoidal varying currents), tracking systems and feedforward control designed to counter open ocean currents, in addition to integral control. The outer loop controller uses Line-of-Site (LOS)

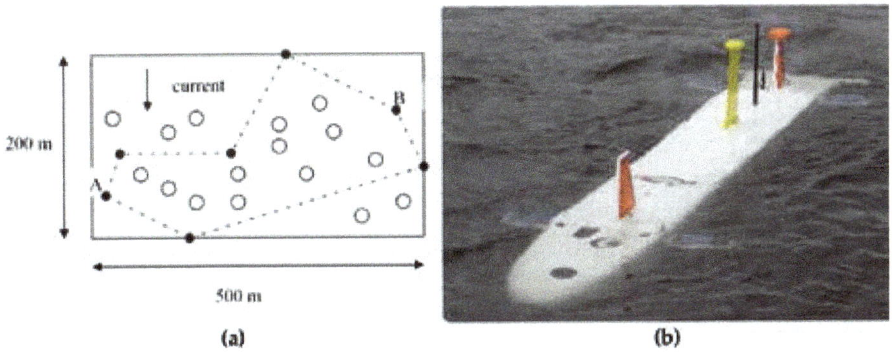

Figure 1. Submersible vehicle sample and notional minefield [1]: (a) field of randomly placed submersed mines to be avoided by autonomous vehicle and (b) *Aries* submersible in open ocean.

guidance to provide a heading command to the inner loop. The inner loop controller uses output heading feedback to track heading commands. The vehicle is simulated to traverse the minefield and successfully travels no closer than 5 m from any mine and arrive within one half meter from the commanded destination autonomously.

2.1. System dynamics

The equations of motion used to simulate the dynamic behavior of the autonomous submersible vehicle in a horizontal plane are listed in Eqs. (1)–(4). All variables in these equations are assumed to be in nondimensional form with respect to the vehicle length (7.3′) and constant forward speed (~3 ft./s). The vehicle weighs 435 lbs. and is neutrally buoyant. Time is nondimensionalized such that 1 s represents the time it takes to travel one vehicle length (**Figure 2**).

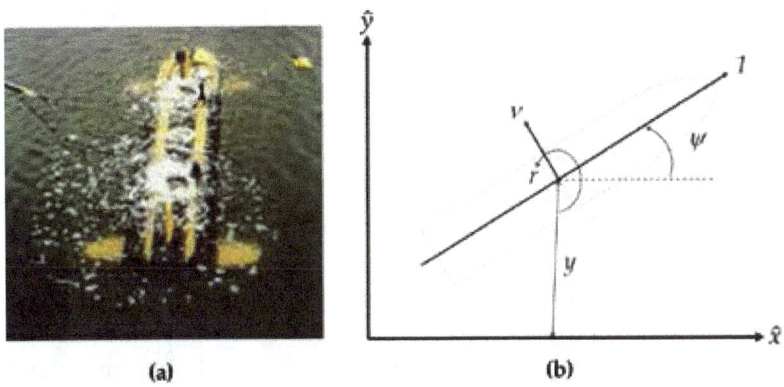

Figure 2. Vehicle geometry and reference axes: (a) *Phoenix* in open ocean [1] and (b) vehicle geometry and reference axis.

$$(m - Y_{\acute{v}})\acute{v} - (Y_{\acute{r}} - mx_G)\acute{r} = Y_{\acute{r}}v + (Y_r - m)r + Y_{\delta_s}\delta_s + Y_{\delta_b}\delta_b \tag{1}$$

$$(mx_G - N_{\acute{v}})\acute{v} - (N_{\acute{r}} - I_z)\acute{r} = N_v v + (N_r - mx_G)r + N_{\delta_s}\delta_s + N_{\delta_b}\delta_b \tag{2}$$

$$\acute{\psi} = r \tag{3}$$

$$\acute{y} = sin\psi + vcos\psi \tag{4}$$

In addition to the following dependent equation $\acute{x} = cos\psi - vsin\psi$ (5)

where

	The variable are	The constants are	Y_{δ_s} = 0.01241.
v	lateral (sway) velocity	m = 0.0358	Y_{δ_b} = 0.01241
r	turning rate (yaw)	I_z = 0.0022	$N\acute{r}$ = – 0.00047
ψ	heading angle (degrees)	x_G = 0.0014	$N\acute{v}$ = – 0.00178
y	lateral deviation (cross-track error)	$Y\acute{r}$ = – 0.00178	N_r = – 0.00390
δ_s	stern rudder deflection	$Y\acute{v}$ = – 0.03430	N_v = – 0.00769
δ_b	bow rudder deflection	Y_r = 0.01187	N_{δ_s} = – 0.0047
		Y_v = – 0.10700	N_{δ_b} = 0.0035

The constant definitions in the mass m, mass moment of inertia with respect to a vertical axis that passes through the vehicle's geometric center (amidships) I_z, position of the vehicle's center of gravity (measured positive forward of amidships) x_G, with the remaining terms referred to as the hydrodynamic coefficients. These constants are all presented in non-dimensional form.

Defining the state vector $\{x\} \equiv \{v \quad r \quad \psi \quad y\}^T$ and the control $\{u\} \equiv \{\delta_s \quad \delta_b\}^T$ and assuming small angles, the dynamics expressed in Eqs. (1)–(4) may be expressed in state space form as $\{\acute{x}\} = [A]\{x\} + [B][u]$ where

$$[A] = \begin{bmatrix} -1.4776 & -0.3083 & 0 & 0 \\ -1.8673 & -1.2682 & 0 & 0 \\ 0 & 1 & 0 & 0 \\ 1 & 0 & 1 & 0 \end{bmatrix} \quad [B] = \begin{bmatrix} 0.2271 & 0.1454 \\ -1.9159 & 1.2112 \\ 0 & 0 \\ 0 & 0 \end{bmatrix} \tag{6}$$

The system may also be expressed in a transfer function ratio of outputs divided by inputs in Laplace form using Eq. (7) where observer matrix [C] is merely a proper identity matrix to this point of the manuscript. Eq. (7) yields two transfer function relationships between each of the two possible rudder inputs as seen in Eqs. (8) and (9). Notice that both transfer functions have poles and zeros at the origin, while pole-zero cancelation is possible in the case of the stern rudder. On the other hand, even after pole-zero cancelation in the bow rudder Eq. (9), there remains an open loop pole at the origin that must be dealt with during control design, since it represents a potentially unstable element (at the very least, in the instance where the estimated constants are exactly correct, and these equations of motion exactly describe the system, an oscillatory element exists that will not decay). Nonetheless, the dynamics accord to nature. Consider trying to steer a row-boat using the rear rudder. It is much more stable than trying to steer the rowboat using a rudder in the front. This analogy applies to the submersible vehicle and is verified in these results.

$$G(S) = [C](s[I] - [A])^{-1}[B] \tag{7}$$

$$G(S)|_{\delta_s} \equiv \frac{Y(s)}{\delta_s(s)} = \frac{0.2271s^3 + 0.875s^2}{s^4 + 2.746s^3 + 1.298s^2} = \frac{s^2(0.2271s + 0.875)}{s^2(s^2 + 2.746s + 1.298)} \tag{8}$$

$$G(S)|_{\delta_b} \equiv \frac{Y(s)}{\delta_b(s)} = \frac{1.211s^2 + 1.518s}{s^4 + 2.746s^3 + 1.298s^2} = \frac{s(1.211s + 1.518)}{s^2(s^2 + 2.746s + 1.298)} \tag{9}$$

In **Figure 3**, the uncontrolled system is analyzed by merely performing a circular turn with each (and then both) rudders. The bow and stern rudders alone are each compared to the combined use of both bow and stern rudders. The bow rudder was deflected +15° for about 21 s, while the stern rudder was deflected for −15° for about 11 s. When both rudders were deflected the maneuver was completed in roughly 8 seconds. Two initial conditions for the sway velocity were investigated ($v(0) = 0$ and then $v(0) = \sqrt{8}$). In all cases, the bow rudder alone performed the poorest, with the stern rudder alone performing the turn in a smaller radius and shorter time. Furthermore, the combined use of both rudders resulting in tightest maneuver.

Two simulation methodologies were used to investigate sensitivities to integration method. MATLAB was used with Euler integration, while SIMULINK was used with Runge-Kutta integration with identical timesteps, Δt = 0.1 s. The results were nearly negligible and are displayed in **Table 1**, from which insensitivity to integration approach is established.

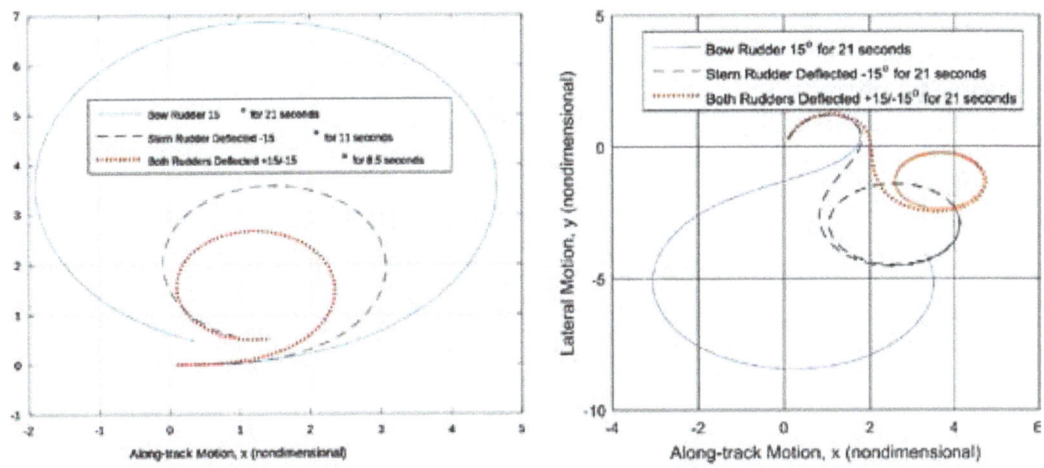

(a) (b)

Figure 3. Analysis of uncontrolled system: comparison of rudder performance: (a) counter-clockwise turn, $v(0) = 0$ and (b) initial sway velocity $v(0) = \sqrt{8}$.

Rudder deflected	Euler: x-distance	Runge-Kutta: x-distance	Euler: y-distance	Runge-Kutta: y-distance
Bow	6.5471	6.5469	6.8647	6.8646
Stern	3.1665	3.1665	3.5768	3.5768
Both	2.4546	2.4546	2.6567	2.6567

[1]Distances calculated to traverse one circular path.

Table 1. Comparison of simulation integration methodology.

2.2. Control law design

In the system analysis, the optimal rudder implementation scheme was determined to be the application of both rudders, where the rudders were slaved to the same maneuver angle magnitude with the opposite sign, i.e. a "scissored-pair" per Eq. (10). In the case where only variable y is to be measured, the new state space formulation of the system equation components are in Eq. (11). Under the assumption of rudders constrained to behave as a scissored-pair the transfer function from rudder input to output y is given by Eq. (12) whose poles and zeros are listed in Eq. (13), with Eq. (14) revealing the system's eigenvalues, noting the values are identical to the location of the poles in accordance with theory. The controllability and observability matrices ($[CO]$ and $[OB]$ respectively) are listed in Eq. (15) (whose matrix product $[OC]$ is in Eq. (16)) verifying these system equations are both controllable and observable, since these matrices are full rank, while the determinant of the controllability matrix is 63.1778, a large value with a small value of the matrix condition number, 13.4513. The nonzero determinant of the controllability matrix proves controllability, but to see how close the system is to being uncontrollable, the matrix condition number proves more useful. These two figures of merit indicate the system equations are highly controllable, and accordingly this manuscript will investigate and compare several options for navigation control: pole placement, linear quadratic optimal control, linear quadratic Gaussian, and time optimal control. The same holds true for observability, and thus linear quadratic Gaussian. The matrix product $[OC]$ is the same for every definition of state variables for the given system.

$$\delta_b = -\delta_s \tag{10}$$

$$[A] = \begin{bmatrix} -1.4776 & -0.3083 & 0 & 0 \\ -1.8673 & -1.2682 & 0 & 0 \\ 0 & 1 & 0 & 0 \\ 1 & 0 & 1 & 0 \end{bmatrix}, [B] = \begin{bmatrix} 0.0816 \\ -3.1271 \\ 0 \\ 0 \end{bmatrix}, \mathrm{C} = [0\,0\,0\,1], \mathrm{D} = [0] \tag{11}$$

$$G(S)|_{\delta} \equiv \frac{Y(s)}{\delta(s)} = \frac{0.08164s^2 - 2.06s - 4.773}{s^4 + 2.746s^3 + 1.298s^2} \tag{12}$$

$$polesat: s = 0, 0, -0.6070, -2.1388; zerosat: s = -6.1279e^{13}, near - 0, near - 0 \tag{13}$$

$$eig(A) = \lambda = 0, 0, -0.6070, -2.1388 \tag{14}$$

$$[CO] = \begin{bmatrix} 0.0816 & 0.8433 & -2.4216 & 5.5544 \\ -3.1271 & 3.8132 & -6.4105 & 12.6514 \\ 0 & -3.1271 & 3.8132 & -6.4105 \\ 0 & 0.0816 & -2.2838 & 1.3916 \end{bmatrix} [OB] = \begin{bmatrix} 0 & 0 & 0 & 1 \\ 1 & 0 & 1 & 0 \\ -1.4776 & 0.6917 & 0 & 0 \\ 0.8917 & -0.4217 & 0 & 0 \end{bmatrix} \quad (15)$$

$$[OC] = \begin{bmatrix} 0 & 0.0816 & -2.838 & 1.3916 \\ 0.0816 & -2.2838 & 1.3916 & -0.8561 \\ -2.2838 & 1.3916 & -0.8561 & 0.5441 \\ 1.3916 & -0.8561 & 0.5441 & -0.3825 \end{bmatrix} \quad (16)$$

Diagonalizing the original system $[A]$ matrix, the spectral decomposition $[T][\Lambda] = [A][T] \rightarrow [\Lambda] = [T]^{-1}[A][T]$ in Eq. (17) may be used to verify a diagonal matrix of eigenvalues $[\Lambda]$, and then write the system of equations in *normal-coordinate form* $\{\acute{x}\prime\} = [A']\{x'\} + [B'][u]; \{y'\} = [C'][x']$ using the following transformation: $[A'] = [\Lambda] = [T]^{-1}[A][T]$, $[B'] = [T]^{-1}[B]$, and $[C'] = [C][T]$ whose results are in Eq. (18).

$$[\Lambda] = \underbrace{\begin{bmatrix} 0.4663 & -0.1074 & 0 & 0 \\ 1 & 0.3033 & 0 & 0 \\ -0.4676 & -0.4996 & 0 & 0 \\ 0.0006 & 1 & 1 & -1 \end{bmatrix}^{-1}}_{T^{-1}} \underbrace{\begin{bmatrix} -1.4776 & -0.3083 & 0 & 0 \\ -1.8673 & -1.2682 & 0 & 0 \\ 0 & 1 & 0 & 0 \\ 1 & 0 & 1 & 0 \end{bmatrix}}_{A} \underbrace{\begin{bmatrix} 0.4663 & -0.1074 & 0 & 0 \\ 1 & 0.3033 & 0 & 0 \\ -0.4676 & -0.4996 & 0 & 0 \\ 0.0006 & 1 & 1 & -1 \end{bmatrix}}_{T} \quad (17)$$

$$[A'] = \begin{bmatrix} -2.1388 & 0 & 0 & 0 \\ 0 & -0.6070 & 0 & 0 \\ 0 & 0 & 0 & 0 \\ 0 & 0 & 0 & 0 \end{bmatrix}, [B'] = \begin{Bmatrix} -1.2502028 \\ -6.1888806 \\ -8.4625e^{7} \\ -8.4625e^{7} \end{Bmatrix}, [C'] = \{0.0006 \quad 1 \quad 1 \quad -1\} \quad (18)$$

$$\{u\}_{baseline} = \{\delta\} = -K_v v - K_r r - K_\psi \psi - K_y y \quad (19)$$

For the pole placement proportional-derivative (PD) controller articulated in Eq. (19), the poles are set to have roughly the same time constant, while avoiding exactly coincident poles. Gains are iterated for various time constants as displayed in **Figure 4**, but the following rule of thumb is asserted as well to quickly achieve performance that closely mimics the performance of linear-quadratic optimal (LQR) gains where the control effort and tracking error are equally weighted in the cost function of the optimization.

RULE OF THUMB: Select unity time-constant t_c to roughly locate closed-loop poles per Eq. (20). Then place other poles at slightly different locations (e.g. $s_p = s_1 \pm 0.01 \forall p$)

$$Pole : s_1 = \frac{1}{t_c} \quad (20)$$

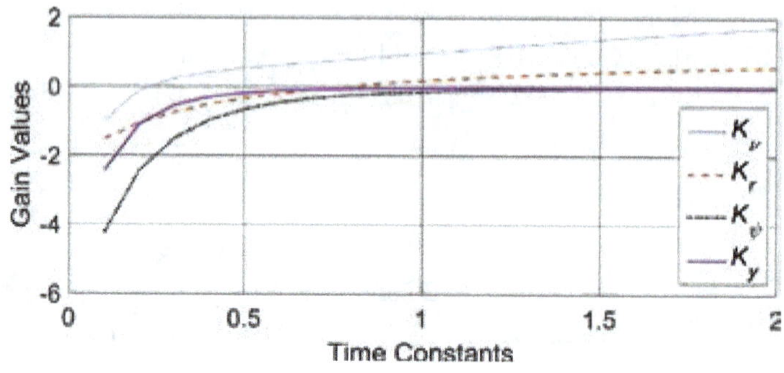

Figure 4. Gain values for each state iterated for various time constants.

The gains achieved using the rule of thumb $K_{R.O.T.}$ = {0.5070 –0.3687 -0.7157 -0.1972} have quite different values compared to the gains calculated through the matrix Riccati equation in the linear-quadratic optimization K_{LQR} = {−0.0939 −1.2043 -2.2138 -1}, but nonetheless the resulting behaviors are indeed very similar.

Next, the initial feedback control design was evaluated in simulations where the ship is initially located off the desired track by one ship's length port side with zero heading, and rudder deflection was limited to 0.4 radians (~23°). Next, another simulation was performed to test an initial heading angle of 30° starboard where the initial $y(0) = 0$. The results are displayed in **Figure 5(a)** and **(b)** respectively. All state variations were plotted in **Figure 4**, highlighting the fact that y converges to zero along with the other states. Furthermore, the results of rudder-limited simulations are displayed in **Figure 6** and **Figure 7** for both scenarios (**Table 2**).

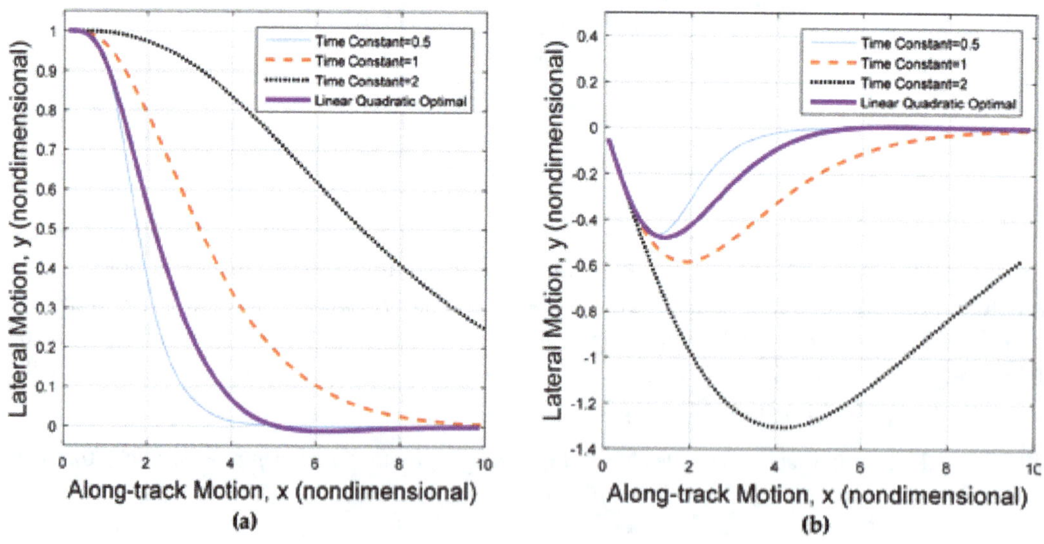

Figure 5. Simulations testing the initial baseline feedback controller in two scenarios: (a) initially one ship's length port side and (b) initial heading 30° starboard.

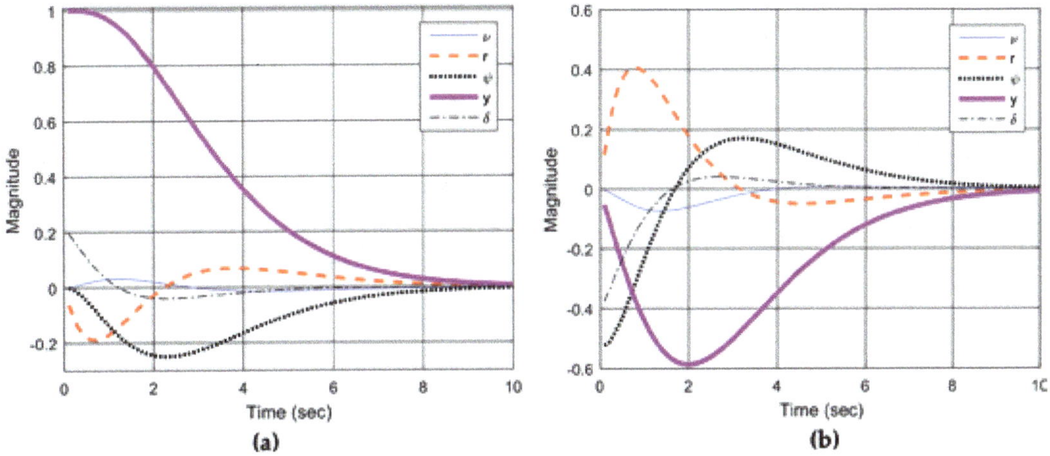

Figure 6. State variations for both scenarios simulated using pole-placement gains via *rule of thumb*: (a) initially one ship's length port side and (b) initial heading 30° starboard.

Time constant	K_v	K_r	K_ψ	K_y
0.5	−1.5135	−1.7005	−5.1508	−3.22524
1	0.5070	−0.3687	−0.7157	−0.1972
2	1.1248	0.2870	−0.0906	−0.0116
LQR	−0.0939	−1.2043	−2.2138	−1

[1]Reminder: state definition $\{x\} \equiv \{v \quad r \quad \psi \quad y\}^T$.

Table 2. Gains for various time constants and also solution to linear quadratic optimization.

2.3. Observer design

To design a state observer, the system must be observable [4], verifiable through examination of the observability matrix [*OB*] per Eq. (21), where [C] = [$v\ r\ \psi\ y$] = [0 1 1 1]. The condition of the observability matrix reveals the degree of observability, and it is defined by the ratio of maximum to minimal singular values.

$$[OB] = \begin{bmatrix} C \\ CA \\ CA^2 \\ \vdots \\ CA^{n-1} \end{bmatrix} \tag{21}$$

2.3.1. *Full-order observer design*

$$\left\{\dot{\hat{x}}\right\} = [A]\{\hat{x}\} + [B][u] + [L](\{y\} - [C]\{\hat{x}\}) \tag{22}$$

$$\left\{\dot{x}\right\} - \left\{\dot{\hat{x}}\right\} = [A]\{x\} - [A]\{\hat{x}\} - [L]([C]\{x\} - [C]\{\hat{x}\}) \tag{23}$$

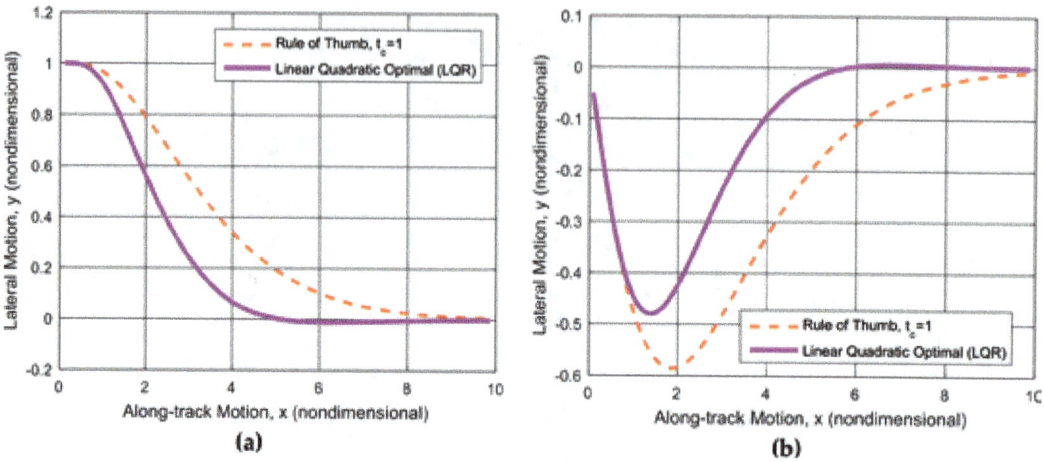

Figure 7. Rudder-limited trajectory track using pole-placement gains via *rule of thumb* and LQR: (a) initially one ship's length port side and (b) initial heading 30° starboard.

$$\left\{\acute{e}\right\} \equiv \left\{\acute{x}\right\} - \left\{\acute{\hat{x}}\right\} = ([A] - [L][C])(\{x\} - \{\hat{x}\}) \tag{24}$$

$$\{\acute{e}\} = ([A] - [L][C])\{e\} \tag{25}$$

Assuming that only v measurements are available, a mathematical model of the estimated system is in Eq. (22) with a full order observer design using the observer error Eq. (23) leading to the error vector in Eq. (24) allowing the re-expression of Eq. (22) as Eq. (25), where the dynamic behavior of the error vector is determined by the eigenvalues of matrix$[A] - [L][C]$, where $[L]$ gains of the observer may be chosen as desired for systems that prove observable, such that the error vector will converge to zero for any stable $[A] - [K_e][C]$. In the following paragraphs, $[L]$ is designed by solving the matrix Ricatti equation leading to linear quadratic optimal gains, and also by solving the *rule of thumb* relationship between gains and time constant as done for the controller gains (**Table 3**).

Figure 8 displays the results of simulations revealing the accuracy of state estimation when $[L]$ is calculated by the *rule of thumb*, where the time constant is chosen to be half ($t_c = 1/2$) the time constant of the controller ($t_c = 1$) and the simulation is initialized with the heading angle 30° off, while **Figure 9** displays the simulation initialized one boat-length starboard position.

Multiple of controller time constant used for observer	Observer gain matrix
$\frac{1}{2}t_c$	$\{-0.7464 \quad 1.8077\ 8.8270 \quad 5.1942\}^T$
$10t_c$	$10^3 * \{-1.5909 \quad -3.4121\ 1.5953 \quad -0.0020\}^T$

[1]Reminder: $\frac{1}{2}t_c$ is used in subsequent simulations.

Table 3. Full-order observer gains designed by *rule of thumb* for various time constants as multiple of controller time constant, t_c.

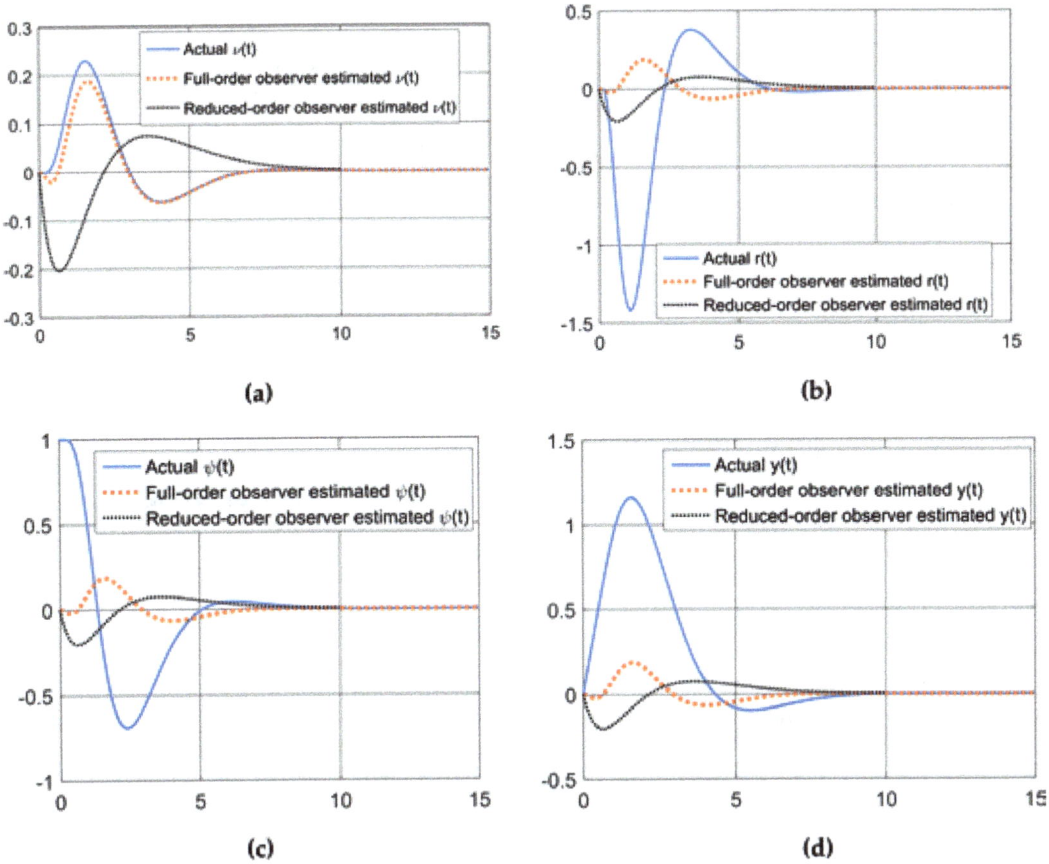

Figure 8. Simulations starting 30° off heading gains via *rule of thumb* state observer gains, (a) true and estimated sway velocity, $v(t)$, (b) true and estimated turning rate, $r(t)$, (c) true and estimated heading angle, $\psi(t)$, (d) true and estimated cross track, $y(t)$.

2.3.2. *Reduced-order observer design*

Assuming that some measurements are available from sensors, this paragraph describes the possible iterations and reveals states that are relatively more important to measure with sensors. Four possible output matrices are used to investigate observability. Four options for output matrices $[C]_i$ for i = 1…4 result in four reduced-order observers $[OB]_i$ for i = 1…4 are detailed in Eqs. (26)–(29). Output matrix $[C]_1$ produces an observability matrix $[OB]_1$ with rank = 4 (observable) and determinant not nearly equal to zero. Output matrix $[C]_2$ produces an observability matrix $[OB]_2$ with rank = 4 (observable) and determinant not nearly equal to zero. Output matrix $[C]_3$ produces an observability matrix $[OB]_3$with rank = 4 (observable) and determinant nearly equal to zero. The matrix condition number is very high indicating the system is barely observable. Output matrix $[C]_4$ produces an observability matrix $[OB]_4$ with rank = 3 (not observable) and determinant equal to zero with a matrix condition number equal to infinity. This means if all other states are measured by sensors, it is not possible to use an observer (even an optimal observer) to determine lateral deviation (cross-track error), y. It is a

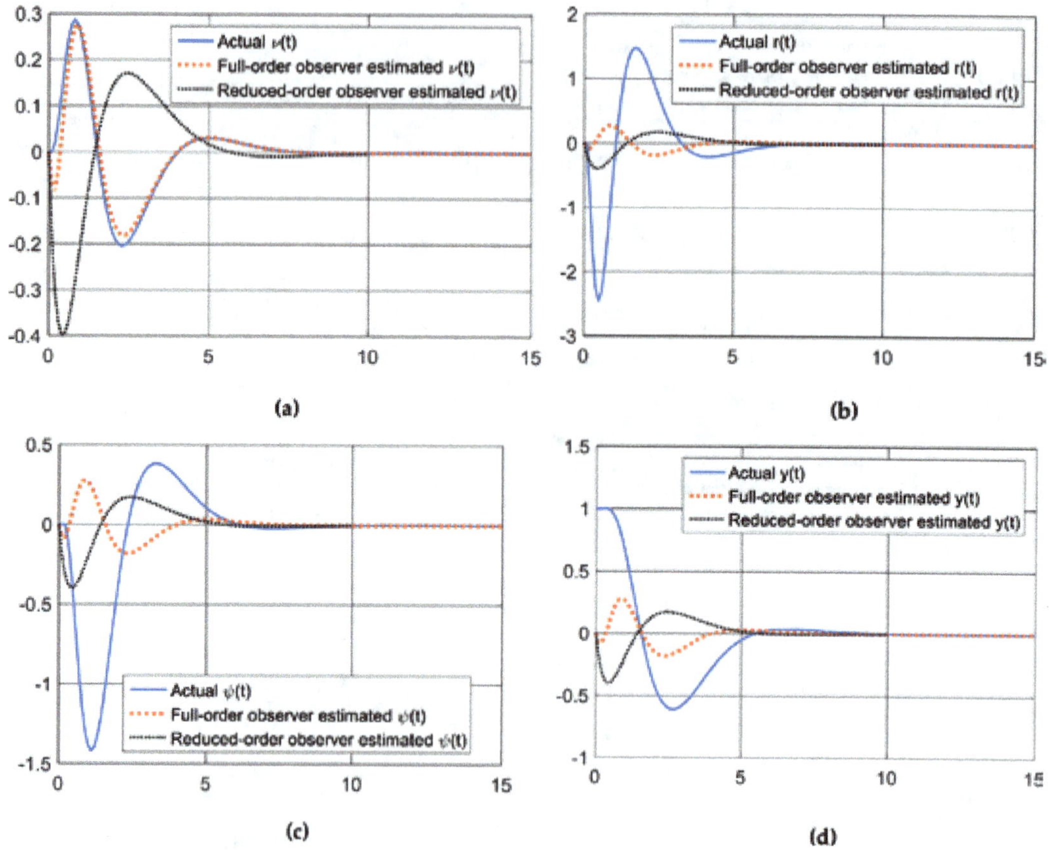

Figure 9. Simulations starting 1 boat-length starboard with gains via *rule of thumb*, (a) true and estimated sway velocity, $v(t)$, (b) true and estimated turning rate, $r(t)$, (c) true and estimated heading angle, $\psi(t)$, (d) true and estimated cross track, $y(t)$.

key state to measure with sensors. The sensor combinations that include y are observable. Using every other sensor, (except y) results in a system that is not observable. Furthermore, measuring y alone results in a barely observable system.

$$[C]_1 = \begin{Bmatrix} \nu \\ r \\ \psi \\ y \end{Bmatrix} = \begin{Bmatrix} 0 \\ 1 \\ 1 \\ 1 \end{Bmatrix} \rightarrow [OB]_1 = \begin{bmatrix} C \\ CA \\ CA^2 \\ \vdots \\ CA^{n-1} \end{bmatrix} = \begin{bmatrix} 0 & 1 & 1 & 1 \\ -0.8673 & -0.2682 & 1 & 0 \\ 1.7823 & 1.6074 & 0 & 0 \\ -5.6352 & -2.5879 & 0 & 0 \end{bmatrix} \quad (26)$$

$$[C]_2 = \begin{Bmatrix} \nu \\ r \\ \psi \\ y \end{Bmatrix} = \begin{Bmatrix} 0 \\ 1 \\ 0 \\ 1 \end{Bmatrix} \rightarrow [OB]_2 = \begin{bmatrix} C \\ CA \\ CA^2 \\ \vdots \\ CA^{n-1} \end{bmatrix} = \begin{bmatrix} 0 & 1 & 0 & 1 \\ -0.8673 & -1.2682 & 1 & 0 \\ 3.6496 & 2.8756 & 0 & 0 \\ -10.7624 & -4.7717 & 0 & 0 \end{bmatrix} \quad (27)$$

$$[C]_2 = \begin{Bmatrix} v \\ r \\ \psi \\ y \end{Bmatrix} = \begin{Bmatrix} 0 \\ 0 \\ 0 \\ 1 \end{Bmatrix} \rightarrow [OB]_3 = \begin{bmatrix} C \\ CA \\ CA^2 \\ \vdots \\ CA^{n-1} \end{bmatrix} = \begin{bmatrix} 0 & 0 & 0 & 1 \\ 1 & 0 & 1 & 0 \\ -1.4776 & 0.6917 & 0 & 0 \\ 0.8917 & -0.4217 & 0 & 0 \end{bmatrix} \quad (28)$$

$$[C]_2 = \begin{Bmatrix} v \\ r \\ \psi \\ y \end{Bmatrix} = \begin{Bmatrix} 1 \\ 1 \\ 1 \\ 0 \end{Bmatrix} \rightarrow [OB]_4 = \begin{bmatrix} C \\ CA \\ CA^2 \\ \vdots \\ CA^{n-1} \end{bmatrix} = \begin{bmatrix} 1 & 1 & 1 & 0 \\ -3.3450 & -0.5764 & 1 & 0 \\ 6.90190 & 1.7621 & 0 & 0 \\ -12.1843 & -4.0900 & 0 & 0 \end{bmatrix} \quad (29)$$

Assuming y is to be measured by a sensor, **Table 4** reveals that measuring v in addition to y produces the most observable system, and is recommended for designing reduced-order observers. The drawback is measuring v requires a Doppler sonar, which may not always be available. If all states are measureable except v the resulting reduced-order observer merely estimates v using gains on the measureable states displayed in **Table 5**. **Figure 10** reveals very good estimation of v when all other states are sensed, and this estimated value of v was fed to the motion controller in addition to the measured states (the poorly estimated states were neglected instead favoring the more-accurate measurements). State convergence to zero is achieved in the instance of state initialization 30° off-heading. **Figure 11** displays similar results for the instance of state initialization one boat-length starboard.

Sensors used to measure states	Observability matrix condition number
y and v	8.8456
y and r	21.1306
y and ψ	31.2919

[1]Reminder: high condition number is less observable system.

Table 4. Observability matrix condition number for options to supplement y measurement.

Multiple of controller time constant used for observer	Observer gain matrix
$\frac{1}{10}t_c$	$\{-0.2174 \quad 0 \quad 0.1164\}^T$
$2t_c$	$\{0.4069 \quad 0 \quad -0.2179\}^T$
$10t_c$	$\{0.5941 \quad 0 \quad -0.3182\}^T$

[1]Relatively faster $\frac{1}{10}t_c$ is used in subsequent simulations.

Table 5. Reduced-order observer gains designed by *rule of thumb* for various time constants as multiple of controller time constant, t_c.

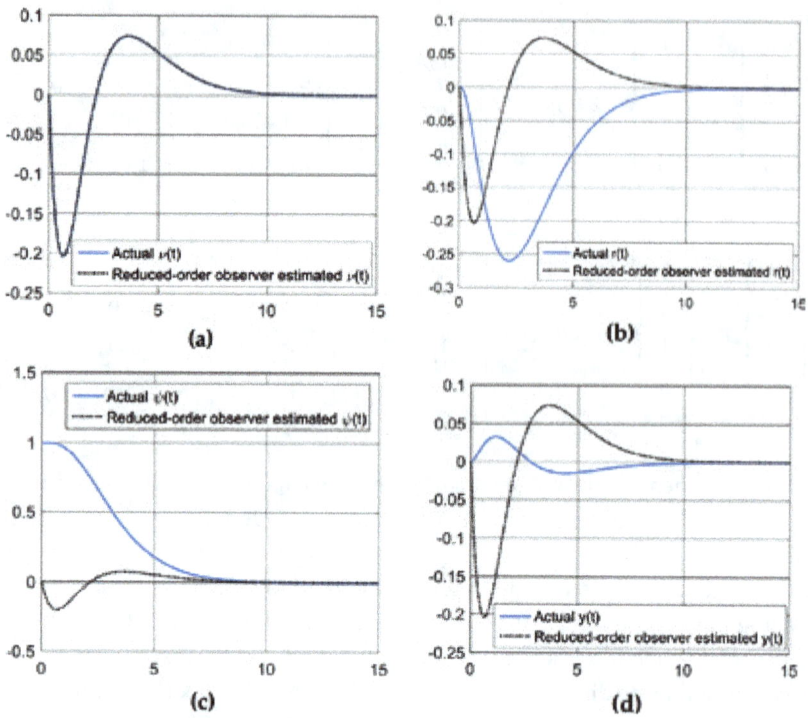

Figure 10. Simulations starting 30° off heading gains via *rule of thumb* reduced-order state observer gains: (a) true and estimated sway velocity, *v(t)* versus time (seconds), (b) true and estimated turning rate, *r(t)* versus time (seconds), (c) true and estimated heading angle, *ψ(t)* versus time (seconds), (d) true and estimated cross track, *y(t)* versus time (seconds).

2.3.3. *Gain margin and phase margin*

Figure 12 compares the loop gains of the system with and without a compensator via gain margin and phase margin with full-state feedback, while **Figure 13** displays the loop gains when output-feedback via observers is used. Each has relative strengths. Full state (theoretical) feedback yields infinite gain margin, yet relatively lower phase margin (usually consider more important of the two), while output feedback (real-world) yields good (but lesser) gain margin with increased phase margin.

2.4. Tracking systems and feedforward control in the presence of constant disturbance currents

This section evolves the earlier developed system equations and performance analysis by adding non-quiescent conditions, in particular introduction of a lateral underwater ocean current with an absolute velocity, v_0, requiring a modification of the system equations to add the lateral current to Eq. (4) resulting in Eq. (30).

$$\acute{y} = sin\psi + vcos\psi + v_0 \tag{30}$$

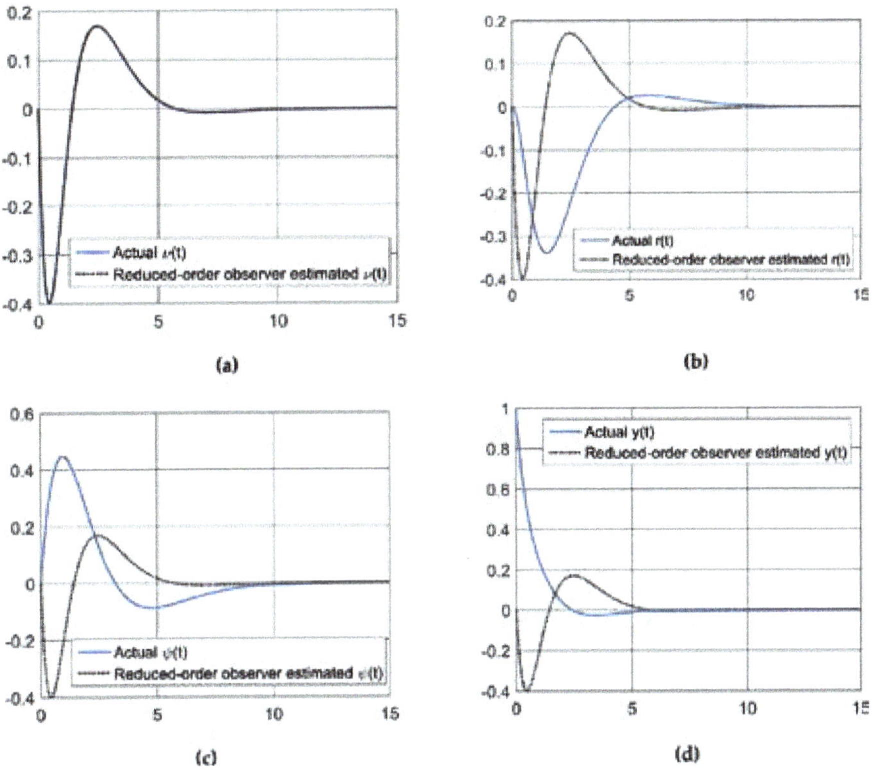

Figure 11. Simulations starting one boat-length starboard with gains via *rule of thumb* reduced-order state observer gains: (a) true and estimated sway velocity, *v(t)* versus time (seconds), (b) true and estimated turning rate, *r(t)* versus time (seconds), (c) true and estimated heading angle, *ψ(t)* versus time (seconds), (d) true and estimated cross track, *y(t)* versus time (seconds).

2.4.1. Analysis of disturbed system in ocean currents via state equations and simulations

Using the controller (Eq. (19)) and the modified system equations where Eq. (4) is replaced by Eq. (30), and applying the final value theorem: $f(t)_{t\to\infty}sF(s)_{s\to 0}$, a steady state value $1/\omega + 1$ has some variable quantity added to unity for various v_0. Thus, steady-state errors exist in all cases with such disturbances, which are verified by simulations depicted in **Figure 14** using gain values from the *rule of thumb* (ROT) for unity time constant. The steady-state errors are directly proportional to the disturbance magnitude. **Figure 15** displays max rudder deflection for the maximal lateral ocean current in the study (to verify the control design continues to remain less than 0.4 radians) where we learned any current greater than 0.4 cannot be eliminated; therefore we next investigate feedforward control and integral control.

2.4.2. Elimination of steady-state error using feedforward control

Modify the control law to $\{u\}_{feedforward} = \{\delta\} = -K_1 v - K_2 r - K_3 \psi - K_4 y - K_0$ in order to eliminate the steady-state error, where K_0 is chosen to insure zero steady-state error, where the feedback gains are chosen by the *rule of thumb* (**Figures 16** and **17**).

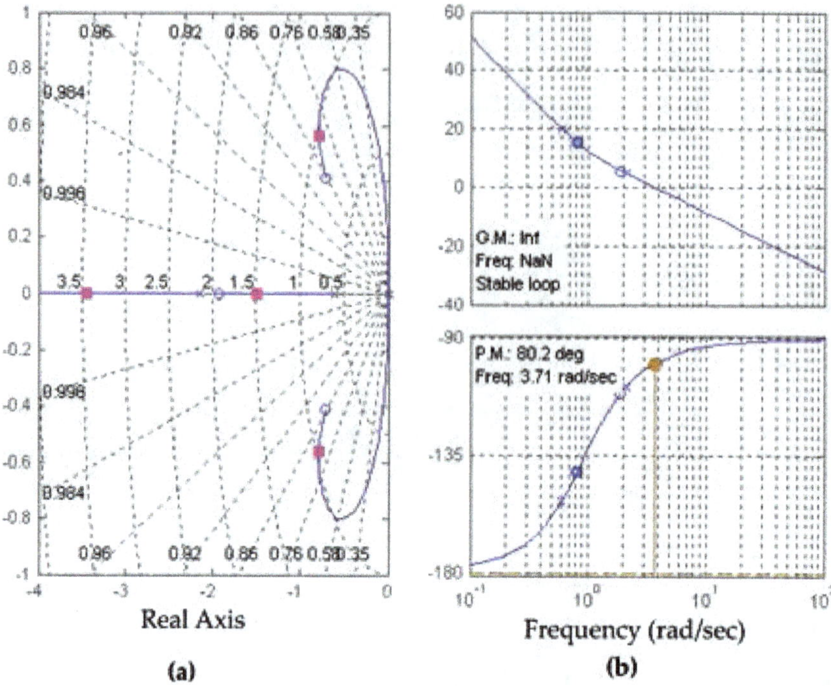

Figure 12. Infinite gain margin and 80.2° phase margin using full state feedback via full-ordered observer with rule of thumb controller gains: (a) root locus real Axis, (b) bode plot frequency (rad/sec).

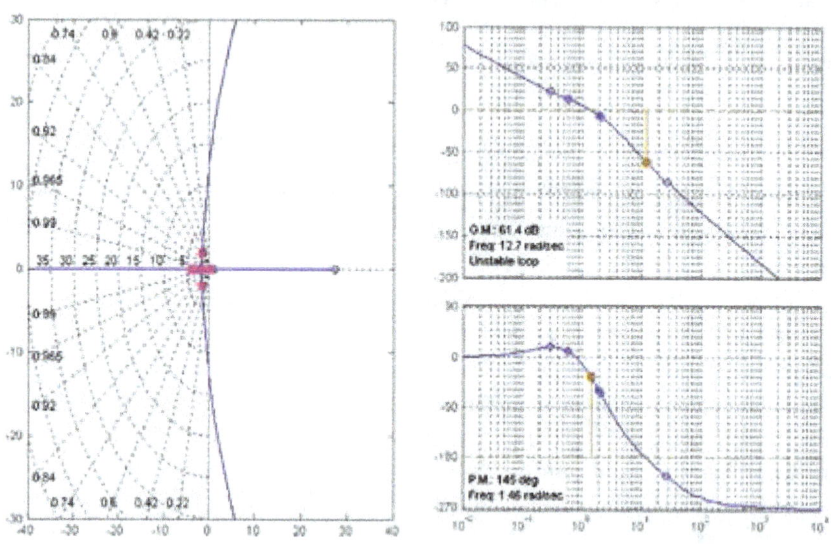

Figure 13. 61.4° gain margin and 145° phase margin using reduced-order observer (both rule of thumb gains for half-controller $t_c = 0.5$, and compensator with rule of thumb gains ($t_c = 1$), (a) root locus, real axis, (b) bode plot, frequency (rad/sec).

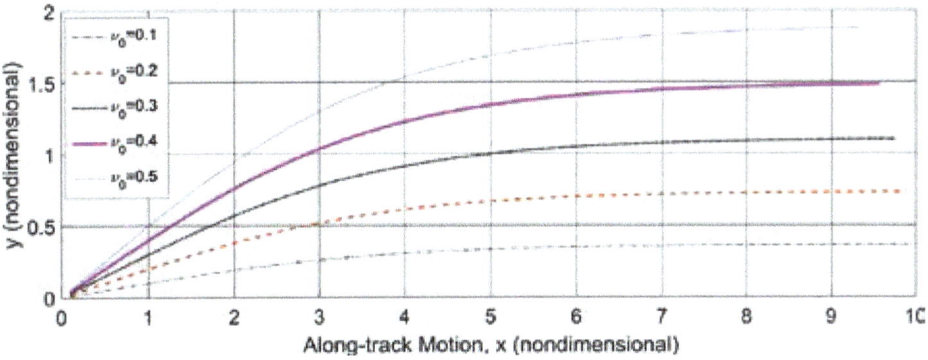

Figure 14. Steady-state position error for various lateral underwater ocean currents.

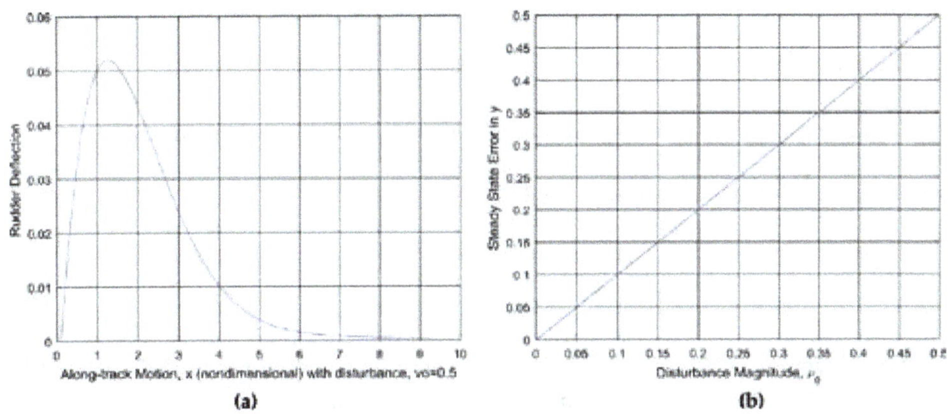

Figure 15. Feedback alone unable to counter constant lateral underwater ocean currents, (a) rudder deflection, $\upsilon_0 = 0.5$, (b) steady state error vs. υ_0.

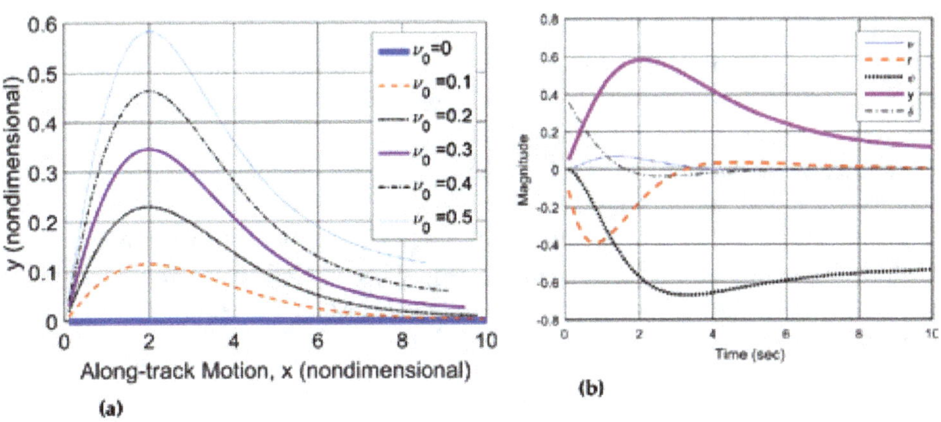

Figure 16. Feedforward element included to counter constant lateral underwater ocean currents, (a) rudder deflection, $\upsilon_0 = 0.5$, (b) all states when $\upsilon_0 = 0.5$.

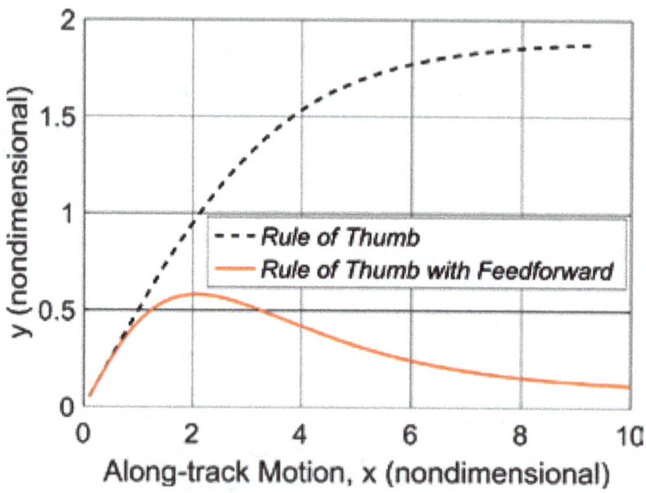

Figure 17. Comparison: feedback control with and without feedforward ($v_0 = 0.5$).

2.5. Disturbance estimation with reduced-order observer and integral control

Section 2.4 demonstrated feedforward control effectively countered the disturbance currents, but the current was presumed to be known. In to truly be effective, the reduced order observer is next augmented to include estimation of the unknown disturbance current velocity $\hat{v}_c$, where the observer now estimates the disturbance current velocity, the lateral sway velocity, v, the lateral deviation (cross-track error), y, and the heading angle ψ. **Figure 18a** and **b** display the estimates of the unknown current for two current velocity conditions: $\hat{v}_{c_1} = v_{est_1} = 0.1$and $\hat{v}_{c_2} = v_{est_2} = 0.5$ respectively, while **Figure 18c** and **d** display the y and ψ states for each current velocity conditions. Notice how large rudder deflections modify the heading angle to the command-tracking value which counters the disturbance current (sometimes referred to as "crabbing"), and after establishing the crab heading angle, the rudder deflection goes towards zero, illustrating the effectiveness of command tracking.

Figure 19 displays all the states versus time in seconds and also the trajectory when a worst-case unknown disturbance current $v_c = -0.5$ is applied and estimated by the reduced-order observer where the observer gains are solutions to the linear quadratic Gaussian optimization. Meanwhile **Figure 20** displays the results in cases utilizing command tracking with reduced order observer and with command: $\psi = -0.5$ and sinusoidal disturbance current $v_{c_0} = A\sin(0.1\ t)$ but no disturbance estimation or feedforward, while **Figure 21** uses disturbance estimation and feedforward and rule of thumb gains. Lastly, **Figure 22** displays the performance of reduced-order observers, which is especially useful in instances of limited at-sea computational capabilities.

2.6. Waypoint guidance

A simple line-of-sight guidance routine was employed based on fixing waypoints through a minefield in order to navigate to a specified point and safely return home. The coordinates are

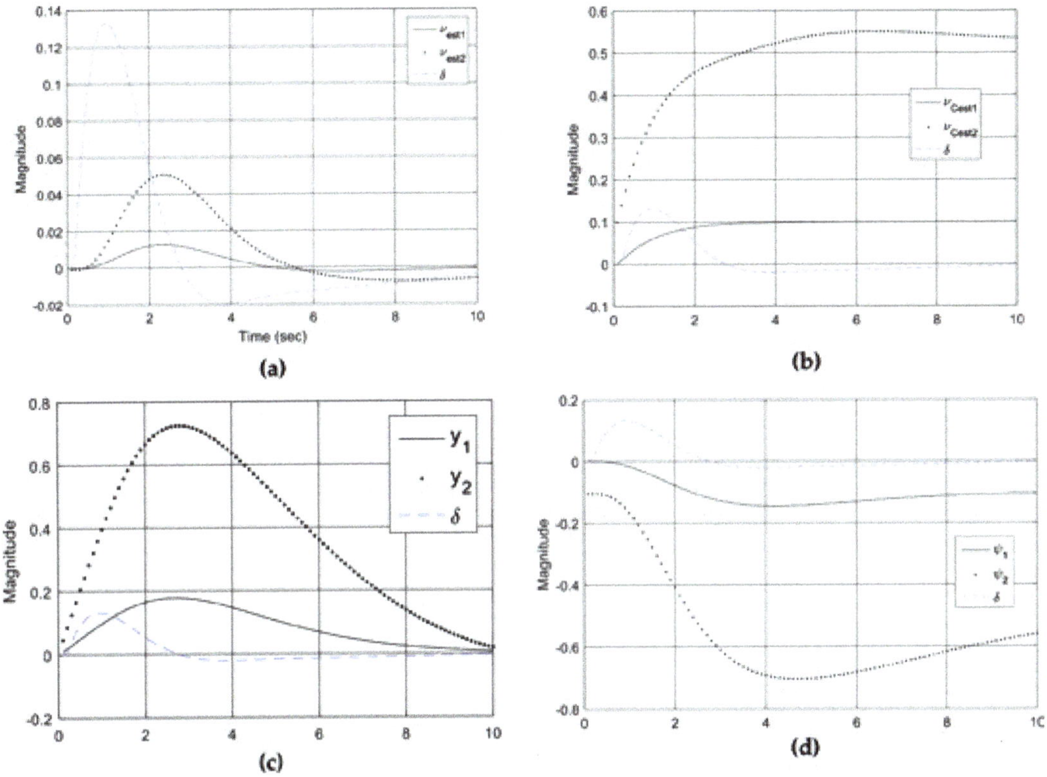

Figure 18. Reduced-order observer state estimates versus time (seconds) for two disturbance currents $v_{c_0} = [0.1 \quad 0.5]$, where Δ is the rudder deflection using these estimates when the worst-case disturbance current is applied. (a) Sway velocity, (b) disturbance current, (c) lateral deviation (cross-track error), (d) heading angle.

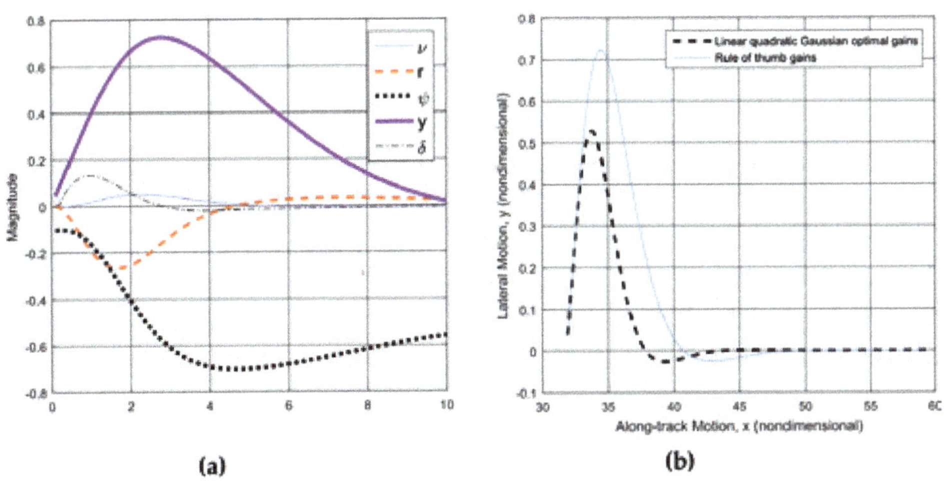

Figure 19. Performance with disturbance estimation and command tracking using LQR and *rule of thumb* gains in reduced order observer, and command tracking to $\psi = -0.5$ amidst constant disturbance current $v_c = 0.5$., (a) states, (b) trajectory.

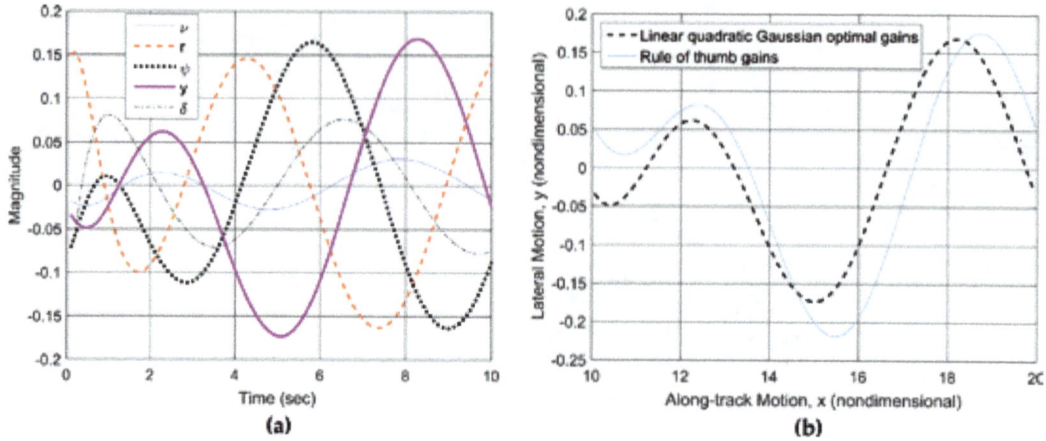

Figure 20. Utilization of command tracking with reduced order observer, with command: $\psi = -0.5$ and sinusoidal disturbance current $v_{c_0} = Asin(0.1t)$ but no disturbance estimation or feedforward, (a) all states vs. time (seconds), (b) trajectory.

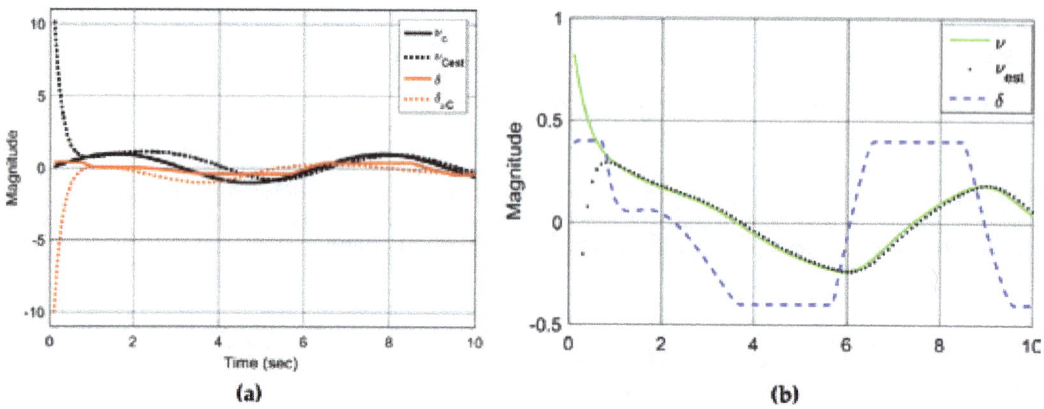

Figure 21. Utilization of command tracking with reduced order observer, with command: ψ = −0.5 and sinusoidal disturbance current $v_{c_0} = Asin(0.1t)$ and disturbance estimation and feedforward and rule of thumb gains, (a), (b).

fed to a logic determining when to turn per Eq. (31), where d is the distance to the waypoint, and the heading command was autonomously calculated per Eq. (32).

$$Turnif : \sqrt{(x_c - x)^2 + (y_c - y)^2} \le d \tag{31}$$

$$\psi_{command} = K\, tan^{-1}\left(\frac{y_c - y}{x_c - x}\right) \tag{32}$$

Particular attention is brought to the inverse tangent calculation, since quadrant must be preserved in the calculation, since the vehicle will navigate in 360°.

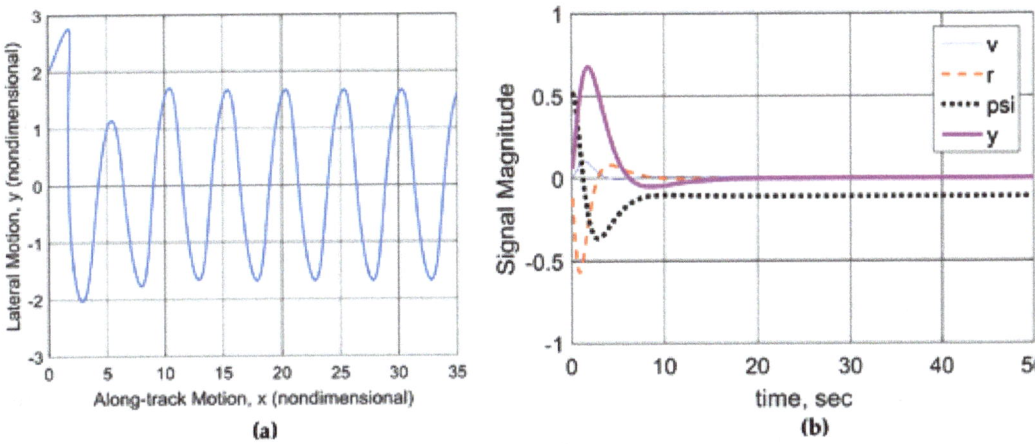

Figure 22. Utilization of command tracking with reduced order observer, with command: $\psi = -0.5$ and sinusoidal disturbance current $v_{c_0} = Asin(0.1t)$, (a) with disturbance estimation (and feedforward), reduced order observer, (b) with integral control but no disturbance estimation or feedforward.

3. Results

The following paragraphs mirror Section 2. Above to provide a concise and precise description of the experimental results, their interpretation as well as the experimental conclusions that can be drawn in each sub-topic introduced and developed so far. Some new development naturally follows in the paragraphs of results, in response to the lessons learned.

3.1. System dynamics

Some basics lessons come from a brief analysis of the uncontrolled system dynamics. The open loop plant equations are potentially unstable (at least persistently oscillatory) with respect to only the bow rudder, while the relationship can be stable with respect to the stern rudder alone. *Can be stable* is exaggerated to emphasize the presence of pole-zero cancelation, which is an unwise practice (especially in this instance with both poles and zeros at the origin on the stability boundary) unless the estimates for the constants in the system equations are very well known. The analysis of the dynamics also revealed the bow rudder was least relatively-effective at maneuvering alone when compared to the stern rudder, however the bow rudder does enhance vehicle maneuverability when used together with the stern rudder as a "scissored-pair" where the sign of the maneuver angle is opposite for each rudder. This "scissored-pair" constraint simplified the MIMO control design, allowing the design engineer to treat the system as a SISO design, since one rudder's deflection become a dependent variable constrained to the other rudder's deflection.

3.2. Control law design

Baseline proportional-derivative control designs effectively stabilized the dynamics, but were ineffective in the presence of a constant lateral open ocean current. Gains selected by *rule of thumb* performed similar to the linear-quadratic optimal control designs, so this underwater vehicle control could be designed at sea with rudimentary math in instances when higher level computational abilities are not available. Augmentation of the control in include gains tuned to reject the constant current proved effective, but required the current to be measure to permit the control component to be properly tuned. Furthermore, when the lateral disturbance current had sinusoidal variation, the controller was rendered ineffective rejecting the disturbance.

3.3. Observer design

The submersible vehicle's system equations were verified observable by calculation of a full-ranked observability matrix in Section 2.3. A full sate observer was designed first to permit vehicle control with "full state feedback", yet without directly measuring velocity. Observer gains may be tuned using classical methods in the general spirit of duality between controller and observers. Their dual nature also permits the matrix Riccati equation to produce optimal gains for a linear-quadratic cost function that exclusively emphasizes state estimation error, unlike the controller optimization where the cost function balanced control effort with state error. State observers permit the vehicle operator to have smooth calculated estimates of all states at all times, which proves useful in the event of sensor interruptions or failures, and reduced-ordered observers may be used in instances where computations on-board the vehicle must be limited, for example to minimize computer size, weight, and/or power.

Especially in light of naturally occurring (roughly) sinusoidal variations in ocean current, the system equations were augmented to include the presumed-unknown disturbance as a state.

3.4. Tracking systems and feedforward control in the presence of disturbance currents

Simple feedforward control elements proved effective against known or estimated constant lateral disturbance currents by allowing the vehicle to autonomously perform "set-and-drift" principles where a highly trained helmsman would turn the bow of a ship into a current, but the simple feedforward elements were ineffective at countering currents with sinusoidal variation. In the set and drift principle the heading is de facto non-zero, so the vehicle cannot simultaneously maintain center-pointing while countering the disturbance. If such a requirement were added, designers must decouple the scissored-pair rudder constraint and design the rudder commands separately to simultaneously counter the disturbance while maintaining centerline pointing.

3.5. Disturbance estimation and integral control

Full-ordered observers effectively estimated constant and sinusoidal disturbance currents and proved useful in the control designs for feedforward control, but furthermore reduced-ordered observer were applied in cases where disturbances were forces and moments and feedforward control was not used. Integral control was used instead to drive steady-state error to zero where sufficiently large time-constants were used for the integrator, i.e. the fifth pole in the pole placement control must be less negative than the other poles.

3.6. Fully assembled system demonstration

In light of all these results, a fully assembled control system was used to navigate the proper mathematical models of the *Phoenix* autonomous submersible vehicle through a simulated 200 m × 500 m minefield in the presence of unknown ocean currents. The field was populated randomly with 30+ mines, and vehicle successfully traversed the minefield in the presence of an unknown 0.5 m/s current with a miss distance from the nearest mine not less than 5 m, navigating from the starting point to pass within 0.5 m of a commanded en route point at sea, and then return to the start point. The outer loop controller used line-of-sight guidance to provide heading commands to the inner loop, and the inner loop controller was an output-feedback heading controller. Two control strategies both proved effective: Linear-quadratic Gaussian, and approximate optimal pole-placement by *rule of thumb*. In the linear-quadratic Gaussian case, both the controller gains and observer gains were selected by optimization of the respective matrix Riccati equation. **Figure 23** displays the completed maneuver where each dot displays the location of a randomly placed mine. Full state feedback was achieved with state observers via the certainly equivalence principle and the states were utilized in a proportional-derivative-integral feedback control architecture. Detailed outputs and figures of merit are plotted in **Figures 24–28** including performance of a second transit of the minefield for validation purposes.

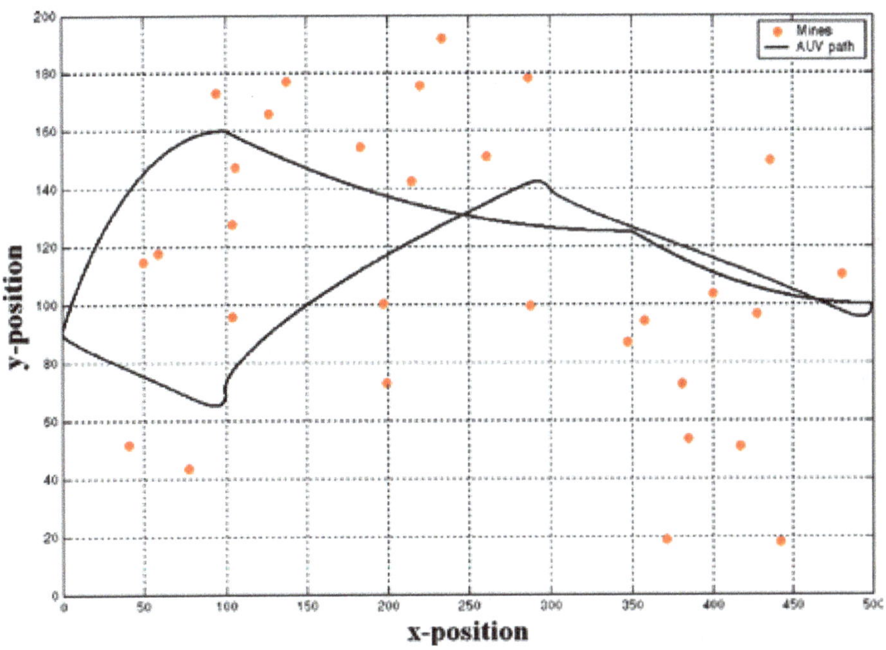

Figure 23. Navigation through simulated field of 30 randomly placed mines in −0.5 m/s current with linear quadratic Gaussian PID controller and full-state observer.

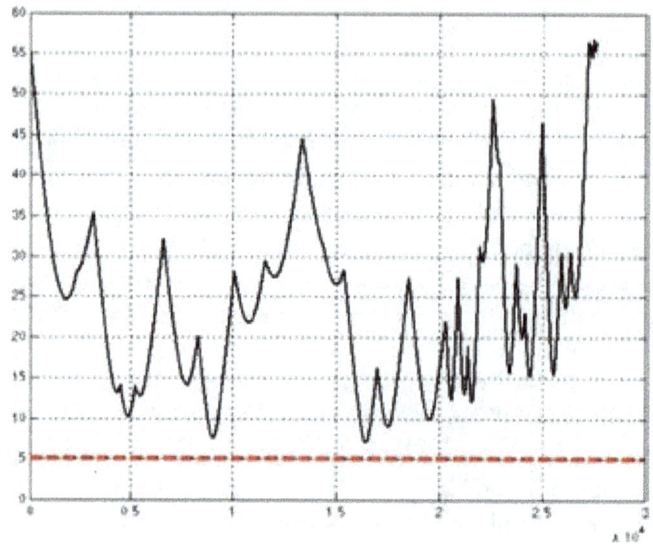

Figure 24. Continuous distance (m) to closest mine with linear quadratic Gaussian optimized *PID* controller and *full-state* observer versus time (s).

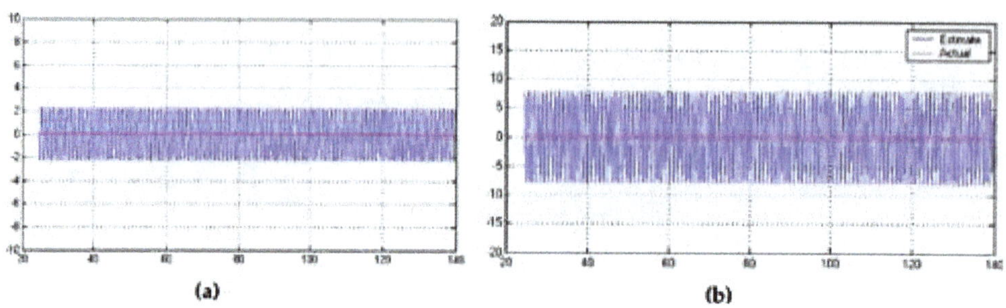

Figure 25. Linear quadratic (Gaussian) optimal observer convergence with actual value in light-pink near zero, while estimates are depicted oscillating in blue, (a) state v, (b) state r.

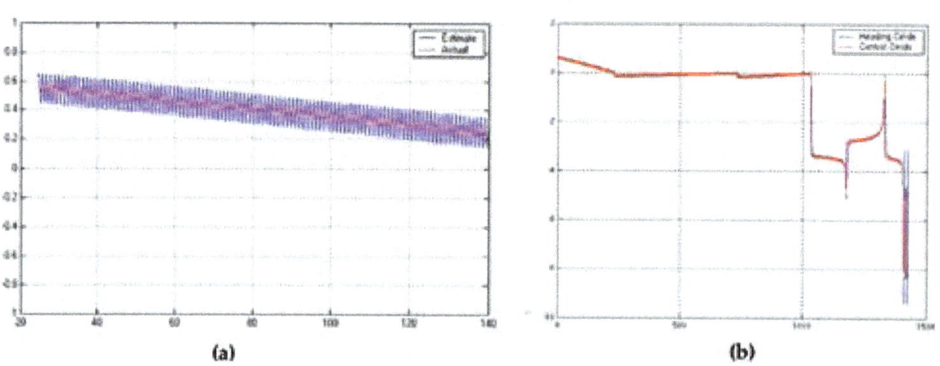

Figure 26. Linear quadratic (Gaussian) optimal observer convergence, (a) state ψ, (b) command tracking (radians) versus time (s).

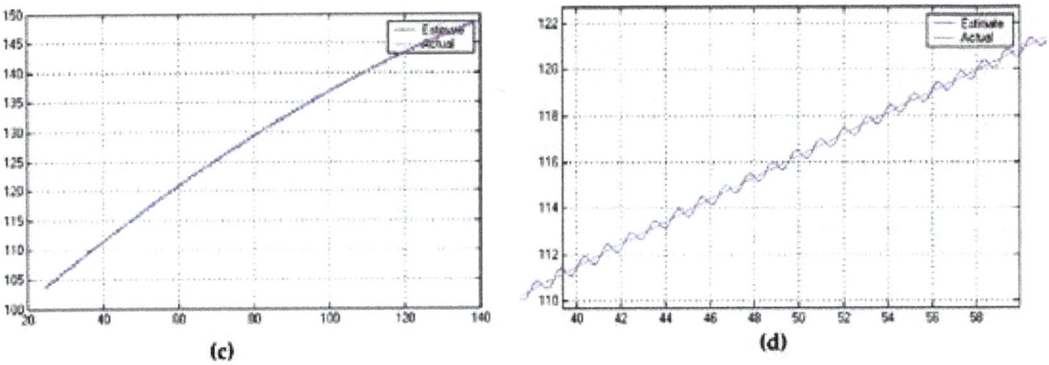

Figure 27. Linear quadratic (Gaussian) optimal observer convergence of y, (c) state y, (d) state.

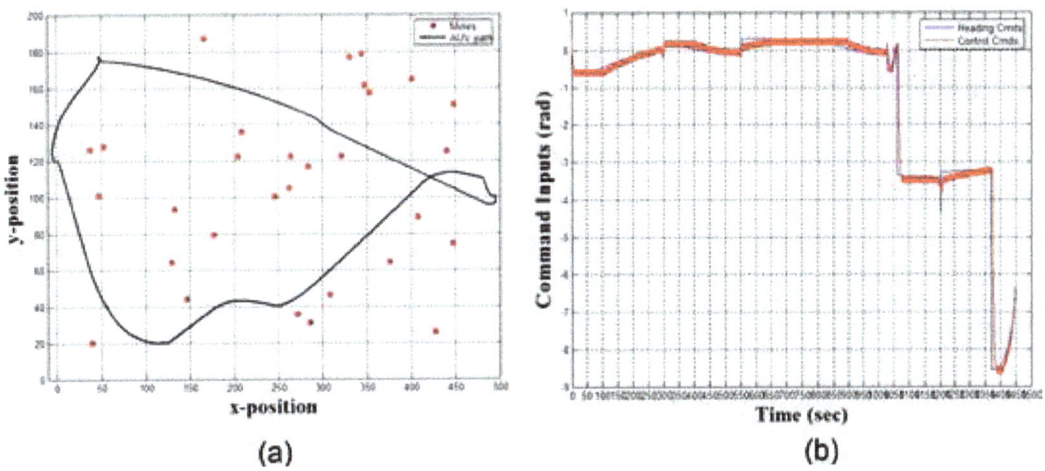

Figure 28. Validation trajectory through simulated field of 30 randomly placed mines in −0.5 m/s current with linear quadratic Gaussian optimized *PI* controller and *reduced-ordered* observer, (a) second trajectory (results validation), (b) heading command tracking.

4. Discussion

The results of this study establish both classical and modern control paradigms to guide autonomous submersible vehicles through obstacles in unknown ocean currents. Both elegant and simplified autonomous controls proved effective, making this technology immediately assessable to low-end technology implementations. The results are consistent with the significant body of literature on motion mechanics in the presence of unknown disturbances with the added complication of restricted path planning due to randomly placed obstacles, where mines were used in this study driving an additional requirement of minimum safe distance for obstacle passage. This consistency with the current literature leads to a natural direction for future research, since recent innovations in nonlinear idealized (and sometimes also adaptive) methods have recently proven to be natural extensions of technology in these fields.

A natural sequel to this manuscript would utilize the aforementioned methods ([26–39] in particular), which comprise nonlinear mathematical amplifications of the linear methods utilized here. The sequel should include an investigation of idealized nonlinear and adaptive methods with a direct comparison to the current state-of-the art including time-optimal control methods.

Acknowledgements

Conceptualization by Distinguished Professor Healey foremost, and then Professor Kevin Bollino and Sands afterwards; methodology by Healey; Software by Bollino and Sands; validation by Professor Isaac Kaminer, Sands, and Bollino; formal analysis by Sands and Bollino; writing-original draft preparation by Sands; writing-review & editing by Sands; research supervision by Kaminer. Authorship has been limited to those who have contributed substantially to the work reported. *Professor Sands is not a tenured faculty member of Stanford University*; rather he is the Associate Dean of the Naval Postgraduate School's Graduate School of Engineering and Applied Sciences. In order to avoid legal jeopardy, Dr. Sands publishes government-funded research under his association with the Naval Postgraduate School, while publishing non-government funded research under his continuing associations with Stanford and Columbia Universities.

Author details

Timothy Sands[1]* and Kevin Bollino[2]

*Address all correspondence to: dr.timsands@stanford.edu

1 Department of Mechanical Engineering, Stanford University, Stanford, USA

2 Department of Electrical and Computer Engineering, George Mason University, Fairfax, USA

References

[1] Naval Postgraduate School website for Consortium for Robotics and Unmanned Systems Education and Research (CRUSER). http://auvac.org/people-organizations/view/241 for Aries AUV [Accessed: 23 May 2018]

[2] Naval Postgraduate School website for Consortium for Robotics and Unmanned Systems Education and Research (CRUSER). https://my.nps.edu/web/cavr/auv for Phoenix AUV [Accessed: 23 May 2018]

[3] Brutzman D, Healey AJ, Marco DB, McGhee RB. The Phoenix autonomous underwater vehicle. In: Kortenkamp B, Murphy, editors. AI Based Mobile Robots. Cambridge, Mass: MIT/AAAI Press; 1998

[4] Ogata K. Modern Control Engineering. 4th ed. New Jersey, USA: Prentice-Hall; 2002. pp. 779-790. ISBN: 0-13-060907-2

[5] Marco DB, Healey AJ. Surge motion parameter identification for the NPS Phoenix AUV. In: Proceedings International Advanced Robotics Program IARP 98; University of South Louisiana; February 18. 1998. pp. 197-210

[6] Kenny T, Sands T. Experimental piezoelectric system identification. Journal of Mechanical Engineering and Automation;**7**:179-195. DOI: 10.5923/j.jmea.20170706.01

[7] Sands T. Nonlinear-adaptive mathematical system identification. Computation. 2017;**5**: 47-59. DOI: 10.3390/computation5040047

[8] Kenny T, Sands T. Experimental sensor characterization. Journal of Space Exploration. 2017;**7**:1. ISSN: 2319-9822

[9] Sands T. Space system identification algorithms. Journal of Space Exploration. 2017;**6**(3): 138. ISSN: 2319-9822

[10] Armani C, Sands T. Analysis, correlation, and estimation for control of material properties. Journal of Mechanical Engineering and Automation. 2018;**2018**(8):7-31. DOI: 10.5923/j.jmea.20180801.02

[11] Wu N, Wu C, Ge T, Yang D, Yang R. Pitch channel control of a REMUS AUV with input saturation and coupling disturbances. Applied Sciences. 2018;**8**:253. DOI: 10.3390/app8020253

[12] He B, Zhang H, Li C, Zhang S, Liang Y, Yan T. Autonomous navigation for autonomous underwater vehicles based on information filters and active sensing. Sensors. 2011;**11**: 10958-10980. DOI: 10.3390/s111110958

[13] Yan Z, Wang L, Zhang W, Zhou J, Wang M. Polar grid navigation algorithm for unmanned underwater vehicles. Sensors. 2017;**17**:1599. DOI: 10.3390/s17071599

[14] Wang J, Li B, Chen L, Li L. A novel detection method for underwater moving targets by measuring their ELF emissions with inductive sensors. Sensors. 1734;**2017**:17. DOI: 10.3390/s17081734

[15] Eren F, Pe'eri S, Thein M-W, Rzhanov Y, Celikkol B, Robinson Swift M. Position, orientation and velocity detection of unmanned underwater vehicles (UUVs) using an optical detector array. Sensors. 2017;**17**:1741. DOI: 10.3390/s17081741

[16] Zhang W, Wei S, Teng Y, Zhang J, Wang X, Yan Z. Dynamic obstacle avoidance for unmanned underwater vehicles based on an improved velocity obstacle method. Sensors. 2017;**17**:2742. DOI: 10.3390/s17122742

[17] Kim JJ, Agrawal B, Sands T. 2H singularity free momentum generation with non-redundant control moment gyroscopes. Proceedings of the IEEE Conference on Decision and Control (CDC). 2006:1551-1556. DOI: 10.1109/CDC.2006.377310

[18] Kim JJ, Agrawal B, Sands T. Control moment gyroscope singularity reduction via decoupled control. Proceedings of the IEEE Southeastcon. 2009:1551-1556. DOI: 10.1109/SECON.2009.5174111

[19] Sands T, Kim JJ, Agrawal B. Nonredundant single-gimbaled control moment gyroscopes. Journal of Guidance Control and Dynamics. 2012;**35**:578-587. DOI: 10.2514/1.53538

[20] Kim JJ, Agrawal B, Sands T. Experiments in control of rotational mechanics. International Journal of Automation, Control and Intelligent Systems. 2016;**2**:9-22. ISSN: 2381-7534

[21] Agrawal B, Kim JJ, Sands T. Method and Apparatus for Singularity Avoidance for Control Moment Gyroscope (CMG) Systems without Using Null Motion. US Patent 9,567,112; 2017

[22] Agrawal B, Kim JJ, Sands T. Singularity penetration with unit delay (SPUD). Mathematics. 2018;**6**:23-38. DOI: 10.3390/math6020023

[23] Lu D, Chu JH, Cheng B, Sands T. Developments in angular momentum exchange. International Journal of Aerospace Sciences. 2018;**6**(1):1-7. DOI: 10.5923/j.aerospace.20180601.01

[24] Thornton B, Ura T, Nose Y. Wind-up AUVs Combined energy storage and attitude control using control moment gyros. In: Proceedings of the Oceans. 2007. DOI: 10.1109/OCEANS.2007.4449149

[25] Thornton B, Ura T, Nose Y. Combined energy storage and three-axis attitude control of a gyroscopically actuated AUV. In: Proceedings of the Oceans; 15-18 September. 2008. DOI: 10.1109/OCEANS.2008.5151885

[26] Sands T. Physics-based control methods. In: Adv. Space. Sys. Orb. Det. London: InTech; 2012. DOI: 10.5772/2408

[27] Nakatani S, Sands T. Simulation of spacecraft damage tolerance and adaptive controls. Proceedings of the IEEE Aerospace. 2014:1-16. DOI: 10.1109/AERO.2014.6836260

[28] Nakatani S, Sands T. Autonomous damage recovery in space. International Journal of Automation, Control and Intelligent Systems. 2016;**2**:22-36. ISSN print: 2381-75

[29] Cooper M, Heidlauf P, Sands T. Controlling chaos—Forced van der pol equation. Mathematics. 2017;**5**:70-80. DOI: 10.3390/math5040070

[30] Nakatani S, Sands T. Battle-damage tolerant automatic controls. Electrical and Electronic Engineering. 2018;**8**:10-23. DOI: 10.5923/j.eee.20180801.02

[31] Sands T. Phase lag elimination at all frequencies for full state estimation of spacecraft attitude. Physik Journal. 2017;**3**:1-12. ISSN: 2471-8491

[32] Sands T. Improved magnetic levitation via online disturbance decoupling. Physik Journal. 2015;**1**(3):272-280. ISSN: 2471-8491

[33] Sands T, Kim JJ, Agrawal B. Spacecraft adaptive control evaluation. Infotech@Aerospace, AIAA2012-2476; 19–21 June 2012; Garden Grove, California; 2012

[34] Sands T, Lorenz R. Physics-based automated control of spacecraft. AIAA SPACE 2009, 6625; 14–17 September; Pasadena, CA, USA. DOI: 10.2514/6.2009–6625

[35] Sands T, Kim JJ, Agrawal B. Improved Hamiltonian adaptive control of spacecraft. IEEE Aerospace Conference; 7–14 March; 1–10 Big Sky, MT, USA. 2009. DOI: 10.1109/AERO.2009.4839565

[36] Sands T, Kim JJ, Agrawal BN. Spacecraft fine tracking pointing using adaptive control. Proc. of SPIE Acquisition, Tracking, Pointing and Laser Systems Technologies XXI, Defense Security Symposium; Orlando, FL, April 9–13; Vol. 6569, 656907; 2007. DOI: 10.1117/12.720694

[37] Sands T. Fine pointing of military spacecraft [Naval Postgraduate School Doctoral Dissertation]. Monterey, California; 2007

[38] Smeresky B, Rizzo A, Sands T. Kinematics in the Information Age. Mathematics 2018. Accepted

[39] Sands T, Bollino K, Kaminer I, Healey A. Autonomous Minimum Safe Distance Maintenance from Submersed Obstacles in Ocean Currents. Journal of Marine Science and Engineering. 2018;**6**:98

10

Robust Accelerating Control for Consistent Node Dynamics in a Platoon of CAVs

Feng Gao, Shengbo Eben Li and Keqiang Li

Abstract

Driving as a platoon has potential to significantly benefit traffic capacity and safety. To generate more identical dynamics of nodes for a platoon of automated connected vehicles (CAVs), this chapter presents a robust acceleration controller using a multiple model control structure. The large uncertainties of node dynamics are divided into small ones using multiple uncertain models, and accordingly multiple robust controllers are designed. According to the errors between current node and multiple models, a scheduling logic is proposed, which automatically selects the most appropriate candidate controller into loop. Even under relatively large plant uncertainties, this method can offer consistent and approximately linear dynamics, which simplifies the synthesis of upper level platoon controller. This method is validated by comparative simulations with a sliding model controller and a fixed $\mathbf{H}_\infty$ controller.

Keywords: automated connected vehicles (CAVs), platoon control, acceleration control, robustness, multi-model

1. Introduction

The platoon driving of automated connected vehicles (CAVs) has considerable potential to benefit road traffic, including increasing highway capacity, less fuel/energy consumption and fewer accidents [1]. The R&D of CAVs has been accelerated with increasing usage of wireless communication in road transportation, such as dedicated short range communications (DSRC). Pioneering studies on how to control a platoon of CAVs can date back to 1990s, and as pointed out by Hedrick et al. , the control topics of a platoon can be divided into two tasks [2, 3]: (1) to implement control of platoon formation, stabilization and dissolution; and (2) to carry out controls for throttle/brake actuators of each vehicle [4]. These naturally lead to a hierarchical control structure, including an upper level controller and a lower level controller [5, 6]. The upper one is to retain safe and string stable operation, whereas the lower one is to track the desired acceleration by determining throttle/brake commands.

The upper level control of a platoon of CAVs has been investigated extensively. An earlier work done by Shladover [2] introduced many known control topics, among which the most famous is the concept of string stability. The string stability ensures that range errors decrease as propagating along downstream [7]. Stankovic et al. [8] proposed a decentralized overlapping control law by using the inclusion principle, which decomposes the original system into multiple ones by an appropriate input/state expansion. Up to now, many other upper level control topics have already been explored, including the influence of spacing policies, information flow topologies, time delay and data loss of wireless communications, etc.

The lower level controller determines the commands for throttle and/or brake actuators. The lower level controller, together with vehicle itself, actually plays the role of node dynamics for upper level control. Many research efforts have been attempted on acceleration control in the past decades, but still few gives emphasis on the request of platoon level automation. Most platoon control relies on one critical assumption that the node dynamics are homogeneous and approximately linear. Then, the node dynamics can be described by simple models, e.g. double-integrator [9, 10] and three-order model [3, 7, 8, 11]. This requires that the behaviour of acceleration control is rather accurate and consistent, which is difficult to be achieved. One is because the salient non-linearities in powertrain dynamics, both traditional [12, 13] and hybridized [14], and any linearization, will lead to errors; the other is that such uncertainties as parametric variations and external disturbances significantly affect the consistence of control behaviour.

One of the major issues of acceleration control is how to deal with non-linearities and uncertainties. The majority to handle non-linearities are to linearize powertrain dynamics, including exact linearization [15, 16], Taylor linearization [17] and inverse model compensation [12, 18]. Fritz and Schiehlen [15, 16] use the exact linearization technique to normalize node dynamics for synthesis of cruising control. After linearization, a pole placement controller was employed to control the exactly linearized states. The Taylor expansion approach has been used by Hunt et al. [17] to approximate the powertrain dynamics at equilibrium points. The gain-scheduling technique was then used to conquer the discrepancy caused by linearization. The inverse model compensation is widely used in engineering practice, for example [12] and [19]. This method is implemented by neglecting the powertrain dynamics. For the uncertainties, the majority rely on robust control techniques, including sliding model control (SMC) [19], $\mathbf{H}_\infty$ control [20, 21], adaptive control [22–24], fuzzy control [25, 26], etc. Considering parametric variations, an adaptive SMC was designed by Swaroop et al. [19] by adding an on-line estimator for vehicle parameters, such as mass, aerodynamic drag coefficient and rolling resistance. Higashimata and Adachi [20] and Yamamura and Seto [21] designed a Model Matching Controller (MMC) based controller for headway control. This design used an $\mathbf{H}_\infty$ controller as feedback and a forward compensator for a faster response. Xu and Ioannou [23] approximated vehicle dynamics to be a first-order transfer function at equilibrium points, and then the Lyapunov approach was used to design an adaptive thriller controller for tracking control of vehicle speed. Keneth et al (2008) designed an adaptive proportional-integral (PI) controller for robust tracking control in resistance to parametric variations. The adaptive law is designed by using the gradient algorithm [24]. The aforementioned robust controllers are useful to resist small errors and disturbances in vehicle longitudinal dynamics, but might not be always effective for large uncertainties. Moreover, the use of adaptive mechanism is only able to resist slowly varying uncertainties, but difficult to respond fast varying disturbances, e.g. instantaneous wind.

2. Node dynamic model for control

This chapter proposes a robust acceleration control method for consistent node dynamics in a platoon of CAVs. This design is able to offer more consistent and approximately linear node dynamics for upper level control of platoons even under large uncertainties, including vehicle parametric variation, varying road slop and strong environmental wind. The controlled node in the platoon is a passenger car with a 1.6 L gasoline engine, a 4-speed automatic transmission, two driving and two driven wheels, as well as a hydraulic braking system. **Figure 1** presents the powertrain dynamics. Its inputs are the throttle angle α_{thr} and the braking pressure P_{brh}. Its outputs are the longitudinal acceleration a, vehicle velocity v, as well as other measurable variables in the powertrain. When driving, the engine torque is amplified by the automatic transmission, final gear, and then acts on two frontal driving wheels. When braking, the braking torque acts on four wheels to dissipate the kinetic energy of vehicle body.

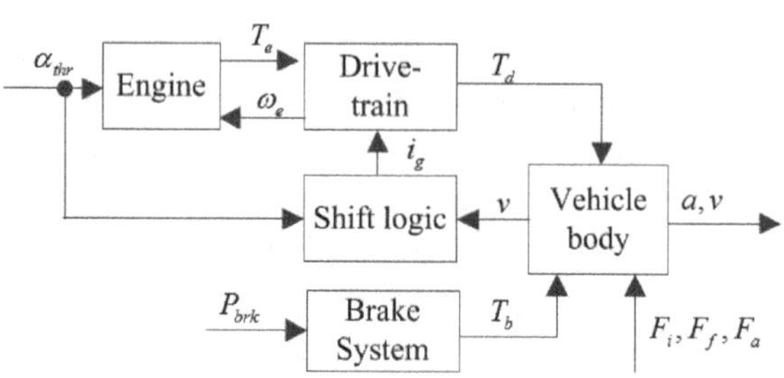

Figure 1. Vehicle longitudinal dynamics.

2.1. Vehicle longitudinal dynamics

For the sake of controller design, it is further assumed that (1) the dynamics of intake manifold and chamber combustion are neglected, and overall engine dynamics are lumped into a first-order inertial transfer function; (2) the vehicle runs on dry alphabet roads with high road-tyre friction, and so the slip of tire is neglected; (3) the vehicle body is considered to be rigid and symmetric, without vertical motion, yaw motion and pitching motion; (4) the hydraulic braking system is simplified to a first-order inertial transfer function without time delay. Then, the mathematical model of vehicle longitudinal dynamics is

$$
\begin{aligned}
&T_{es} = \mathbf{MAP}\left(\omega_e, \alpha_{thr}\right), T_e = \frac{1}{\tau_e s+1} T_{es}, J_e \dot{\omega}_e = T_e - T_p, \\
&T_p = C_{TC}\omega_e^2, T_t = K_{TC}T_p, T_d = \eta_T i_g i_0 T_t, \omega_t = i_g i_0 \frac{v}{r_w}, \\
&M\dot{v} = \frac{T_d}{r_w} - \frac{T_b}{r_w} - F_i - F_a - F_f, T_b = \frac{K_b}{\tau_b s+1} P_{brk}, F_i = Mg \cdot \sin(\varphi), \\
&F_a = \text{sign}\left(v+v_{wind}\right) C_A \left(v+v_{wind}\right)^2, F_f = Mg \cdot f,
\end{aligned} \tag{1}
$$

where $\boldsymbol{\omega_e}$ is the engine speed, $\boldsymbol{T_{es}}$ is the static engine torque, $\boldsymbol{\tau_e}$ is the time constant of engine dynamics, $\mathbf{T_e}$ is the actual engine torque, $\mathbf{MAP}_{(.,.)}$ is a non-linear tabular function representing engine torque characteristics, $\boldsymbol{T_p}$ is the pump torque of torque converter (TC), $\boldsymbol{J_e}$ is the inertia of fly wheel, $\boldsymbol{T_t}$ is the turbine torque of TC, $\boldsymbol{C_{Tc}}$ is the TC capacity coefficient, $\boldsymbol{K_{TC}}$ is the torque ratio of TC, $\boldsymbol{i_g}$ is the gear ratio of transmission, $\boldsymbol{i_o}$ is the ratio of final gear, η_T is the mechanical efficiency of driveline, $\boldsymbol{r_w}$ is the rolling radius of wheels, $\boldsymbol{M}$ is the vehicle mass, $\boldsymbol{T_d}$ is the driving force on wheels, $\boldsymbol{T_b}$ is the braking force on wheels, $\boldsymbol{v}$ is the vehicle speed, $\boldsymbol{F_i}$ is the longitudinal component of vehicle gravity, $\boldsymbol{F_a}$ is the aerodynamic drag, $\boldsymbol{F_f}$ is the rolling resistance, $\boldsymbol{K_b}$ is the total braking gain of four wheels, $\boldsymbol{\tau_b}$ is the time constant of braking system, $\boldsymbol{C_A}$ is the coefficient of aerodynamic drag, g is the gravity coefficient, f is the coefficient of rolling resistance, $\boldsymbol{\phi}$ is the road slope and $\boldsymbol{v_{wind}}$ is the speed of environmental wind. The nominal values of vehicle parameters are shown in **Table 1**.

Symbol	Units	Nominal value
M	Kg	1300
J_e	kg·m^2	0.21
η_T	–	0.89
τ_e	Sec	0.3
i_o	–	4.43
i_g	–	[2.71, 1.44, 1, 0.74]
r_w	M	0.28
K_b	N·m/MPa	1185
τ_b	Sec	0.15
C_A	kg/m	0.2835
f	–	0.02
g	m/s^2	9.81

Table 1. Nominal parameters of vehicle model.

2.2. Inverse vehicle model

One major challenge of acceleration control is the salient non-linearities, including engine static non-linearity, torque converter coupling, discontinuous gear ratio, quadratic aerodynamic drag and the throttle/brake switching. These non-linearities can be compensated by an inverse vehicle model. The inverse models of engine and brake are described by Eqs. (2) and (3), respectively [22, 31]. The design of the inverse model assumes that (i) engine dynamics, torque converter coupling, etc. is neglected; (ii) vehicle runs on dry and flat road with no mass transfer; (iii) the inverse model uses nominal parameters in **Table 1**.

$$T_{\mathrm{edes}} = \frac{r_w}{i_g i_0 \eta_T}\left(Ma_{\mathrm{des}} + C_A v^2 + Mgf\right), \alpha_{\mathrm{thrdes}} = \mathrm{MAP}^{-1}\left(\omega_e, T_{\mathrm{edes}}\right), \tag{2}$$

$$F_{\mathrm{bdes}} = Ma_{\mathrm{des}} + C_A v^2 + Mgf, P_{\mathrm{brkdes}} = \frac{1}{K_b} F_{\mathrm{bdes}}, \tag{3}$$

where $\boldsymbol{a}_{\mathrm{des}}$ is the input for the inverse model, which is the command of acceleration control, $\boldsymbol{T}_{\mathrm{edes}}$, $\boldsymbol{\alpha}_{\mathrm{thrdes}}$, $\boldsymbol{F}_{\mathrm{bdes}}$ and $\boldsymbol{P}_{\mathrm{brkdes}}$ are corresponding intermittent variables or actuator commands. Note that throttle and braking controls cannot be applied simultaneously. A switching logic with a hysteresis layer is required to determine which one is used. The switching line for separation is not simply to be zero, i.e. $\boldsymbol{a}_{\mathrm{des}} = \mathbf{0}$, because the engine braking and the aerodynamic drag are firstly used, and followed by hydraulic braking if necessary. Therefore, the switching line is actually equal to the deceleration when coasting, shown in **Figure 2**. The use of a hysteresis layer is to avoid frequent switching between throttle and brake controls.

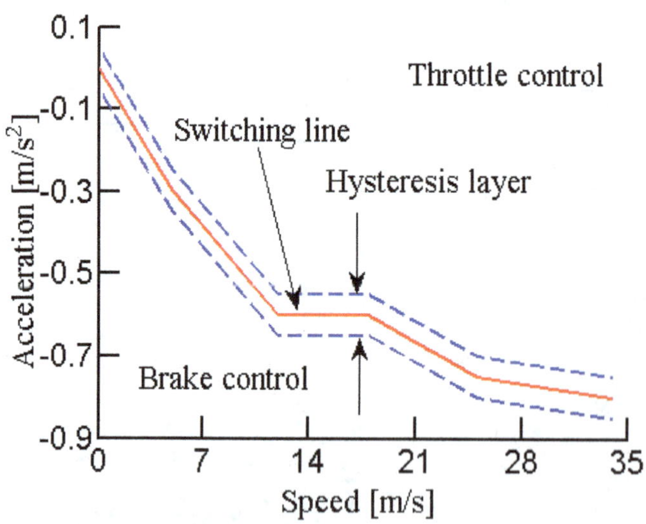

Figure 2. Switching between throttle and brake controls.

3. MMS-based acceleration control

The Multi Model Switching (MMS) control is an efficient way to control plants with large model uncertainties and linearization errors, especially sudden changes in plant dynamics [27–30]. The overall range of plant dynamics is covered by a set of models instead of a single one, and then a scheduling logic switches the most appropriate controller into the control loop. The speed of adaptation and transient performance can be significantly improved by the instantaneous switching among candidate controllers [29, 30]. Another benefit of MMS control is its potential to enclose the input-output behaviours to a required small range. **Figure 3** shows the MMS control structure for vehicle acceleration tracking, where a_{des} and a are the desired and actual longitudinal acceleration respectively, α_{thrdes} and P_{brkdes} are throttle angle and braking pressure respectively, which are the control inputs of a vehicle. It consists of the vehicle itself (**V**), the inverse model (**I**), a supervisor (**S**) and a controller set (**C**). The inverse model **I** is used to compensate for the non-linearities of powertrain; **I** and **V** together constructs the plant for MMS control. The combination of **I** + **V** tends to have large uncertainties, but is divided into small ones under the MMS structure. Such a configuration is able to maintain a more accurate and consistent input–output behaviour even under a large model mismatch.

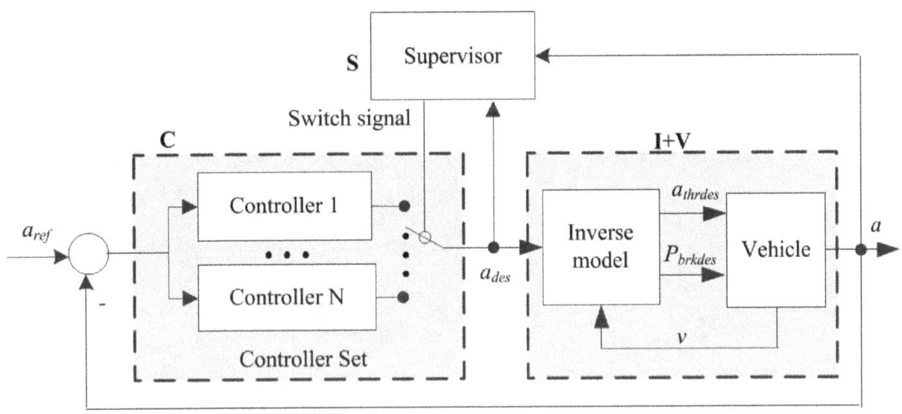

Figure 3. MMS control of vehicle acceleration.

3.1. Model set to separate large uncertainty

For the MMS control, **I** and **V** are combined together to form a new plant, whose input is desired acceleration and output is actual acceleration. Its major uncertainties arise from the change of operating speed, i.e. $v \in \mathbb{R}^1$, the parameter variation, i.e. $\theta = [M, \eta_T, \tau_e, K_b] \in \mathbb{R}^4$. and the external disturbance, i.e. $d = [\varphi, v_{\text{wind}}] \in \mathbb{R}^2$. Their uncertain range is $v \in [v_{\min}, v_{\max}]$, $\theta \in [\theta_{\min}, \theta_{\max}]$ and $d \in [d_{\min}, d_{\max}]$. The main idea is to use multiple linear models, i.e. $P_i(s)$, $i = 1, \cdots, N$, to separate such large uncertainties into small ones, and accordingly design multiple feasible $\mathbf{H}_\infty$ controllers, i.e. $C_i(s)$, $i = 1, \cdots, N$, for each model with smaller uncertainty.

The range of vehicle speed, $v \in \mathbb{R}$, is equally divided into five points, i.e. $\left(v_{\min}, \frac{3v_{\min}+v_{\max}}{4}, \frac{2v_{\min}+2v_{\max}}{4}, \frac{v_{\min}+3v_{\max}}{4}, v_{\max}\right)$, and the range of $\theta \in \mathbb{R}^4$ and $d \in \mathbb{R}^2$ are each separated into three points, i.e. $\left(\theta_{\min}, \frac{\theta_{\min}+\theta_{\max}}{2}, \theta_{\max}\right)$, and $\left(d_{\min}, \frac{d_{\min}+d_{\max}}{2}, d_{\max}\right)$, respectively. Their combination is set as a candidate for model identification. Totally, there are $5 \times 3^{(4+2)} = 3465$ candidate models. The 3465 models can be straightforwardly regarded as a multiple model set. Its shortcoming is that some of these models are quite closed to each other, which naturally leads to many redundant controllers (a waste of computing and storage resources). To reduce the model number, some close models are grouped and covered by an uncertain model. Hence, these 3465 models are clustered into four groups, which are covered with four uncertain models $P_i(s)$, $i=1, \cdots, N$, shown in **Figure 4**.

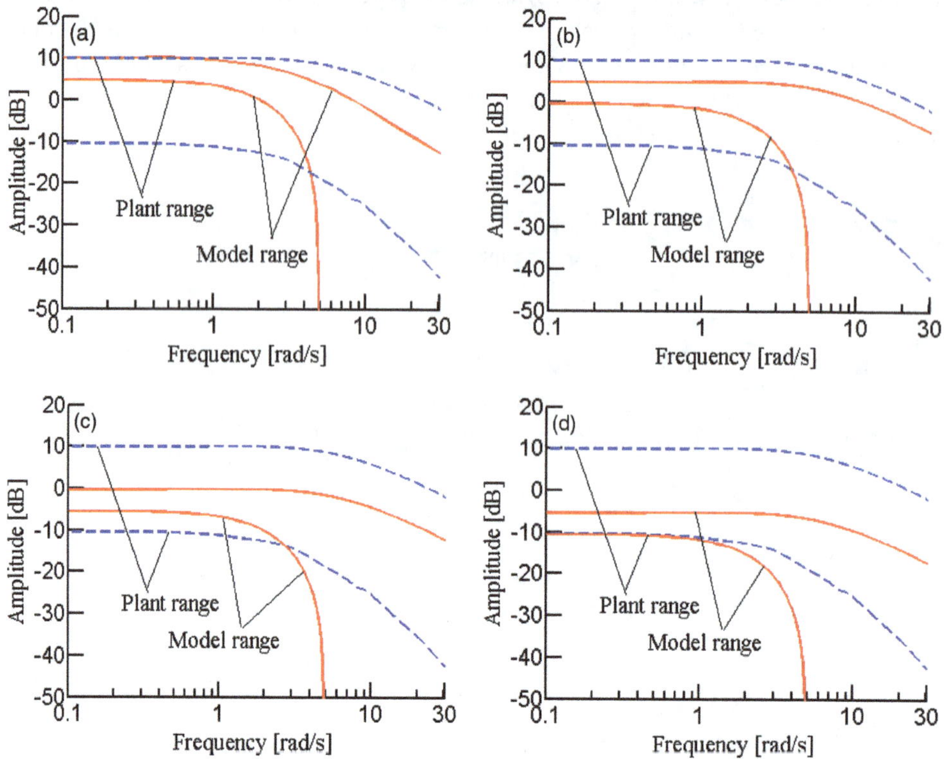

Figure 4. Frequency responses of four linear models. (a) Model $P_1(s)$, (b) Model $P_2(s)$, (c) Model $P_3(s)$ and (d) Model $P_4(s)$.

The model set $\mathbf{P} = \{P_i(s), i=1, \cdots, N\}$ is defined to have identical structure with a multiplicative uncertainty:

$$P_i(s) = G_i(s)\left[1+\Delta_i W(s)\right], G_i(s) = k_{Gi}(s+p_G)^{-1}, i=1\cdots N, \tag{4}$$

where $G_i(s)$ is the nominal models listed in **Table 2**, $W(s)$ is the weight function for uncertainty, Δ_i is the model uncertainty, satisfying:

$$\left\|\Delta_i\right\|_{\infty}^{\delta} < 1, i = 1 \cdots N, \tag{5}$$

where $\|\cdot\|_{\infty}^{\delta}$ is the induced norm of $\mathbf{L}_2^{\delta}$ norm of signals expressed as

$$\left\|\mathbf{x}(t)\right\|_2^{\delta} = \sqrt{\int_0^t e^{-\delta(t-\tau)} \left|\mathbf{x}(t)\right|^2 d\tau}, \tag{6}$$

where $\delta > 0$ is a forgetting factor, and $\mathbf{x}(t)$ is a vector of signals.

No.	1	2	3	4
$G_i(s)$	$\frac{8.15}{s+3.333}$	$\frac{4.5}{s+3.333}$	$\frac{0.75}{s+3.333}$	$\frac{0.22}{s+3.333}$

Table 2. Nominal models of node dynamics.

3.2. Synthesis of the MMS controller

The main idea of this chapter is to use multiple uncertain models to cover overall plant dynamics, and so the large uncertainty is divided into smaller ones. Because the range of dynamics covered by each model is reduced using multiple models, this MMS control can greatly improve both robust stability and tracking performance of vehicle acceleration control. This MMS controller includes a scheduling logic **S**, multiple estimators **E** and multiple controllers **C**. The module **E** is a set of estimators, which is designed from model set **P** and to estimate signals a and z. Note that z is the disturbance signal arising from model uncertainty and it cannot be measured directly. The module **S** represents the scheduling logic. Its task is to calculate and compare the switching index of each model $J_i(i=1, \cdots, N)$, which actually gives a measure of each model uncertainty compared with current vehicle dynamics. **S** chooses the most proper model (with smallest measure) and denoted as σ. The module **C** contains multiple robust controllers, also designed from **P**. The controller whose index equals σ will be switched into loop to control acceleration. The signal a_{ref} is the desired acceleration.

The scheduling logic is critical to the MMS controller, because it evaluates errors between current vehicle dynamics and each model in **P**, and determines which controller should be chosen. The controller index σ is determined by:

$$\sigma = \arg\min_{i=1,\cdots,4} J_i(t). \tag{7}$$

Intuitively, $J_i(t)$ is designed to measure the model uncertainty σ_i, and so the estimator set $\mathbf{E}=\{E_i, i=1, \cdots, N\}$ is designed to indirectly measure σ_i as follows:

$$\hat{z}_i = -\frac{k_{Gi}}{\Lambda(s)} W(s) a_{des}, \hat{a}_i = \frac{k_{Gi}}{\Lambda(s)} a_{des} + \frac{\Lambda(s)-(s+p_G)}{\Lambda(s)} a, i=1,\cdots,4, \tag{8}$$

where $\Lambda(s)$ is the common characteristic polynomial of $\mathbf{E}$, $\hat{a}_i$ and $\hat{z}_i$ are the estimates of a and z using model P_i. It is easy to know that the stability of estimators can be ensured by properly selecting $\Lambda(S)$. Subtracting Eq. (8) with Eq. (4) yields the estimation error of a:

$$e_i = \hat{a}_i - a = -\frac{k_{Gi}}{\Lambda(s)} W(s) \Delta_i a_{des} = \Delta_i \hat{z}_i. \tag{9}$$

Then, the switching index $J_i(t)$ is designed to be

$$J_i(t) = \left(\left\|e_i(t)\right\|_2^{\delta}\right)^2 - \left(\left\|\hat{z}_i(t)\right\|_2^{\delta}\right)^2, i=1,\cdots,4. \tag{10}$$

Since the system gain from $\hat{z}_\sigma(t)$ to $e_\sigma(t)$ can be bounded by $\mathbf{S}$, $\hat{z}_\sigma(t)$ and $e_\sigma(t)$ can be treated as the input and output of an equivalent uncertainty. Considering Eq. (8), $\mathbf{E}$ is rewritten into

$$\begin{gathered} \dot{\mathbf{x}}_E = \mathbf{A}_E \mathbf{x}_E + \mathbf{B}_{E1} a_{des} + \mathbf{B}_{E2} a \\ \hat{a}_i = \mathbf{C}_{E1i} \mathbf{x}_E, \hat{z}_i = \mathbf{C}_{E2i} \mathbf{x}_E, \end{gathered} \tag{11}$$

where A_E, B_{E1}, B_{E2}, C_{E11}, C_{E12}, C_{E13}, C_{E14}, C_{E21}, C_{E22}, C_{E23}, C_{E24} are matrices with proper dimensions. By selecting weighting function as $W_p(s)=(0.1s+1.15)/s$, the required tracking performance becomes:

$$\|q(t)\|_2^{\delta} < \gamma \|a_{\text{ref}}(t)\|_2^{\delta}, \tag{12}$$

where a_{ref} is the reference acceleration, $q = W_{\text{p}}(s)e_{\text{a}}$ and $e_{\text{a}} = a_{\text{ref}} - a,$, which is expected to converge to zero. Substituting $a = \hat{a}_{\sigma} - e_{\sigma}$ and Eq. (12) to Eq. (11), we have,

$$\dot{\mathbf{x}} = \mathbf{A}_{\sigma}\mathbf{x} + \mathbf{B}_1 e_{\sigma} + \mathbf{B}_2 a_{\text{des}} + \mathbf{B}_3 a_{\text{ref}} \tag{13}$$

$$\hat{z}_{\sigma} = \mathbf{C}_{1\sigma}\mathbf{x}, q = \mathbf{C}_{2\sigma}\mathbf{x} + D_{\text{p}}e_{\sigma} + D_{\text{p}}a_{\text{ref}}, e_{\text{a}} = \mathbf{C}_{3\sigma}\mathbf{x} + e_{\sigma} + a_{\text{ref}},$$

where $\mathbf{A}_{\sigma} = \begin{bmatrix} \mathbf{A}_{\text{E}} + \mathbf{B}_{\text{E2}}\mathbf{C}_{\text{E1}\sigma} & 0 \\ -\mathbf{C}_{\text{E1}\sigma} & 0 \end{bmatrix}$, $\mathbf{B}_1 = \begin{bmatrix} -\mathbf{B}_{\text{E2}} \\ -1 \end{bmatrix}$, , $\mathbf{B}_2 = \begin{bmatrix} \mathbf{B}_{\text{E1}} \\ 0 \end{bmatrix}$, $\mathbf{B}_3 = \begin{bmatrix} 0 \\ 1 \end{bmatrix}$, $\mathbf{C}_{1\sigma} = [\mathbf{C}_{\text{E2}\sigma} \;\; 0]$, $\mathbf{C}_{2\sigma} = [-0.1\mathbf{C}_{\text{E1}\sigma} \;\; 1.15]$, $\mathbf{C}_{3\sigma} = [-\mathbf{C}_{\text{E1}\sigma} \;\; 0]$, $\mathbf{D}_{31} = 1$, $\mathbf{x} = \begin{bmatrix} \mathbf{x}_{\text{E}} \\ x_p \end{bmatrix}$ are system matrices with proper dimensions. The required robust controller set can be designed by numerically solving the following LMIs:

$$\begin{bmatrix} \mathbf{A}_{\delta i}\mathbf{P}_1 + \mathbf{P}_1\mathbf{A}_{\delta i}^{\text{T}} + \mathbf{B}_2\tilde{\mathbf{C}}_i + \tilde{\mathbf{C}}_i^{\text{T}}\mathbf{B}_2^{\text{T}} & * & * & * & * & * \\ \left(\tilde{\mathbf{A}}_i^{\text{T}} + \mathbf{A}_{\delta i} + \mathbf{B}_2\tilde{\mathbf{D}}_i\mathbf{C}_{3i}\right)^{\text{T}} & \mathbf{P}_2\mathbf{A}_{\delta i} + \mathbf{A}_{\delta i}^{\text{T}}\mathbf{P}_2 + \tilde{\mathbf{B}}_i\mathbf{C}_{3i} + \mathbf{C}_{3i}^{\text{T}}\tilde{\mathbf{B}}_i^{\text{T}} & * & * & * & * \\ \left(\mathbf{B}_2\tilde{\mathbf{D}}_i + \mathbf{B}_1\right)^{\text{T}} & \left(\mathbf{P}_2\mathbf{B}_1 + \tilde{\mathbf{B}}_i\right)^{\text{T}} & -\beta^2\mathbf{I} & 0 & * & * \\ \left(\mathbf{B}_2\tilde{\mathbf{D}}_i + \mathbf{B}_3\right)^{\text{T}} & \left(\mathbf{P}_2\mathbf{B}_3 + \tilde{\mathbf{B}}_i\right)^{\text{T}} & 0 & -\gamma^2\mathbf{I} & * & * \\ \mathbf{C}_{1i}\mathbf{P}_1 & \mathbf{C}_{1i} & 0 & 0 & \mathbf{I} & 0 \\ \mathbf{C}_{2i}\mathbf{P}_1 & \mathbf{C}_{2i} & D_p & D_p & 0 & \mathbf{I} \end{bmatrix} < 0, i = 1, \cdots, 4, \begin{bmatrix} \mathbf{P}_1 & \mathbf{I} \\ \mathbf{I} & \mathbf{P}_2 \end{bmatrix} > 0, \tag{14}$$

where $\beta < 1$ is a positive constant, $\mathbf{A}_{\delta i} = \mathbf{A}_i + 0.5\delta\mathbf{I}$, symbol "*" represents the symmetrical part. Then the controller set $\mathbf{C}$ is

$$\mathbf{C} = \left\{ K_i : \begin{matrix} \dot{\mathbf{X}}_C = \mathbf{A}_{Ci}\mathbf{X}_C + \mathbf{B}_{Ci}e_a \\ a_{des} = \mathbf{C}_{Ci}\mathbf{X}_C + \mathbf{D}_{Ci}e_a \end{matrix}, i = 1, \cdots, N \right\}. \quad (15)$$

The matrices in Eq. (15) are calculated as:

$$\begin{aligned} \mathbf{D}_{Ci} &= \tilde{\mathbf{D}}_i, \mathbf{C}_{Ci} = \left(\tilde{\mathbf{C}}_i - \mathbf{D}_{Ci}\mathbf{C}_{3i}\mathbf{P}_1 \right)\left(\mathbf{M}^{\mathrm{T}}\right)^{-1}, \mathbf{B}_{Ci} \\ &= \mathbf{N}^{-1}\left(\tilde{\mathbf{B}}_i - \mathbf{P}_2\mathbf{B}_2\mathbf{D}_{Ci} \right), \end{aligned} \quad (16)$$

$$\begin{aligned} \mathbf{A}_{Ci} &= \mathbf{N}^{-1}\left[\tilde{\mathbf{A}}_i - \mathbf{P}_2\left(\mathbf{A}_{\delta i} + \mathbf{B}_2\mathbf{D}_{Ci}\mathbf{C}_{3i}\right)\mathbf{P}_1 \right]\left(\mathbf{M}^{\mathrm{T}}\right)^{-1} \\ &-0.5\delta\mathbf{I} - \mathbf{B}_{Ci}\mathbf{C}_{3i}\mathbf{P}_1\left(\mathbf{M}^{\mathrm{T}}\right)^{-1} - \mathbf{N}^{-1}\mathbf{P}_2\mathbf{B}_2\mathbf{C}_{Ci},\ i = 1, \cdots, 4, \end{aligned}$$

where $\mathbf{M}$ and $\mathbf{N}$ are the singular value decomposition of $\mathbf{I} - \mathbf{P}_1\mathbf{P}_2$. The controller set $\mathbf{C}$, solved by LMIs Eq. (14), is listed as follows:

$$\mathbf{C} = \{K_1(s), K_2(s), K_3(s), K_4(s)\}$$

$$\begin{aligned} K_1(\mathrm{s}) &= \frac{137.1(\mathrm{s}+4.9)(\mathrm{s}+3.133)}{s(\mathrm{s}+41.85)(\mathrm{s}+45.70)},\ K_2(\mathrm{s}) \\ &= \frac{233.4(\mathrm{s}+4.9)(\mathrm{s}+3.133)}{\mathrm{s}(\mathrm{s}+80.06)(\mathrm{s}+21.42)}, \end{aligned} \quad (17)$$

$$\begin{aligned} K_3(\mathrm{s}) &= \frac{573.0(\mathrm{s}+4.9)(\mathrm{s}+3.133)}{\mathrm{s}(\mathrm{s}+29.63)(\mathrm{s}+99.30)},\ K_4(\mathrm{s}) \\ &= \frac{283.4(\mathrm{s}+4.9)(\mathrm{s}+3.133)}{\mathrm{s}(\mathrm{s}+54.15)(\mathrm{s}+19.89)} \end{aligned}$$

4. Simulation results and analyses

To validate the improvements MMS controller for tracking of acceleration, two other controllers are designed, i.e. a sliding mode controller (SMC), and a single $\mathbf{H}_\infty$ controller.

4.1. Design of SMC and $\mathbf{H}_\infty$ controllers

It is known that SMC has high robustness to uncertainties. It is designed based on the nominal model $G_M(s)=0.33/(s+0.33)$. The sliding surface is selected to be

$$e(t)=a-a_{\text{ref}},\ s=\int_0^t e(t)dt+\lambda e(t), \tag{18}$$

where $\lambda>0$. The reaching law is designed to be $\dot{s}=-ks+\eta\text{sgn}(s)$, where $k<0$ and $\eta>0$. Then the sliding mode controller is

$$a_{\text{des}}=\frac{\tau_i}{\lambda k_i}\left[\lambda\dot{a}_{\text{ref}}+\lambda\frac{1}{\tau_i}a-e-ks+\eta\text{sgn}(s)\right] \tag{19}$$

$\mathbf{H}_\infty$ control is another widely used and effective approach to deal with model uncertainties. Here, a model matching control structure is applied to balance between robustness and fastness. The uncertain model of vehicle dynamics used for design of $\mathbf{H}_\infty$ controller is

$$P(s)=\frac{0.3s+1}{0.2s^2+0.6s+1}\left(1+\frac{5.2s+5}{2s+10}\Delta\right),\|\Delta\|_\infty<1 \tag{20}$$

The referenced acceleration response dynamics is $G_M(s)=1/(s+1)$. Then, the feed-forward controller designed by the model matching technique is

$$C_F(s)=\frac{0.2s^2+0.6s+1}{0.3s^2+1.3s+1} \tag{21}$$

The feedback controller designed by the $\mathbf{H}_\infty$ control method is

$$C_B(s)=\frac{6.5493(s+5)(s+6)(s^2+3s+5)}{s(s+10.39)(s+4.74)(s^2+7.049s+14.03)} \tag{22}$$

This $\mathbf{H}_\infty$ controller is numerically solved by the Matlab command *mixsyn*(), with the weighting function $W_p(s)=(0.1s+1.15)s^{-1}$.

4.2. Simulations and analyses

A naturalistic acceleration from real traffic flow is used as the reference acceleration. This naturalistic acceleration profile is from driver experiment data, which lasts around 50 min totally and is shown in **Figure 5**. The maximum and minimum desired acceleration is about 1.1 m/s² and −1.8 m/s², respectively and the vehicle speed varies in the range of 0–33 m/s. This condition can cover a wide range of vehicle dynamics.

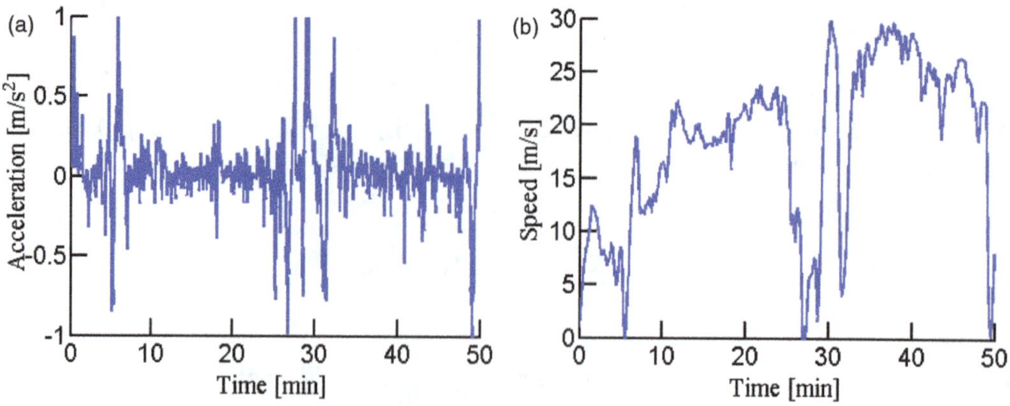

Figure 5. *Reference acceleration* and *speed*. (a) *Acceleration and* (b) *Vehicle* speed.

Two groups of simulations are conducted: (a) nominal condition; and (b) uncertain condition. Under the nominal condition, all vehicle parameters are shown in **Table 1** and there is no road slope and wind. Under the uncertain condition, the disturbed parameters are vehicle mass, road slope and wind. The maximum value of vehicle mass is used, i.e. M = 1600 kg. The disturbance of road slope is a sinusoidal signal:

$$\varphi = \varphi_{\max} \cdot \sin\left(\frac{2\pi}{T_{\mathrm{slope}}} \cdot t\right) \tag{23}$$

where $\varphi_{\max}$ = 5 deg, and T_{slope} = 50 sec. The disturbance of wind is a periodic triangular signal:

$$v_{\mathrm{wind}} = 2\frac{v_{\mathrm{wmax}}}{T_{\mathrm{wind}}}t - v_{\mathrm{wmax}}, t \in \left[0, T_{\mathrm{wind}}\right), \tag{24}$$

where v_{wmax} = 10 m/s, and T_{wind} = 40 sec.

The simulations results are shown in **Figures 6** and **7**, and to show clearly, only the responses from 0 to 500 sec are plotted as a demonstration. From **Figure 6(a)** and (**b**), under the nominal condition, all three controllers can track the reference acceleration accurately. With the uncertainties, it is found that from **Figure 7(a)** and (**b**) the tracking capability of SMC and H_∞ controller decreases obviously while MMS can still ensure acceptable tracking error. Though switching of controller occurs at both nominal and uncertain conditions (**Figure 6(e)** and **Figure 7(e)**), the control input of MMS behaves continuously and there is no obvious sudden change (**Figure 6(c)** and **Figure 7(c)**). Another concern that must be explained is the spike of acceleration shown in **Figure 6(b)** and **Figure 7(b)**. This is mainly caused by the impact of powertrain when gear switches. A more appropriately designed transmission model could improve the gear-shifting quality.

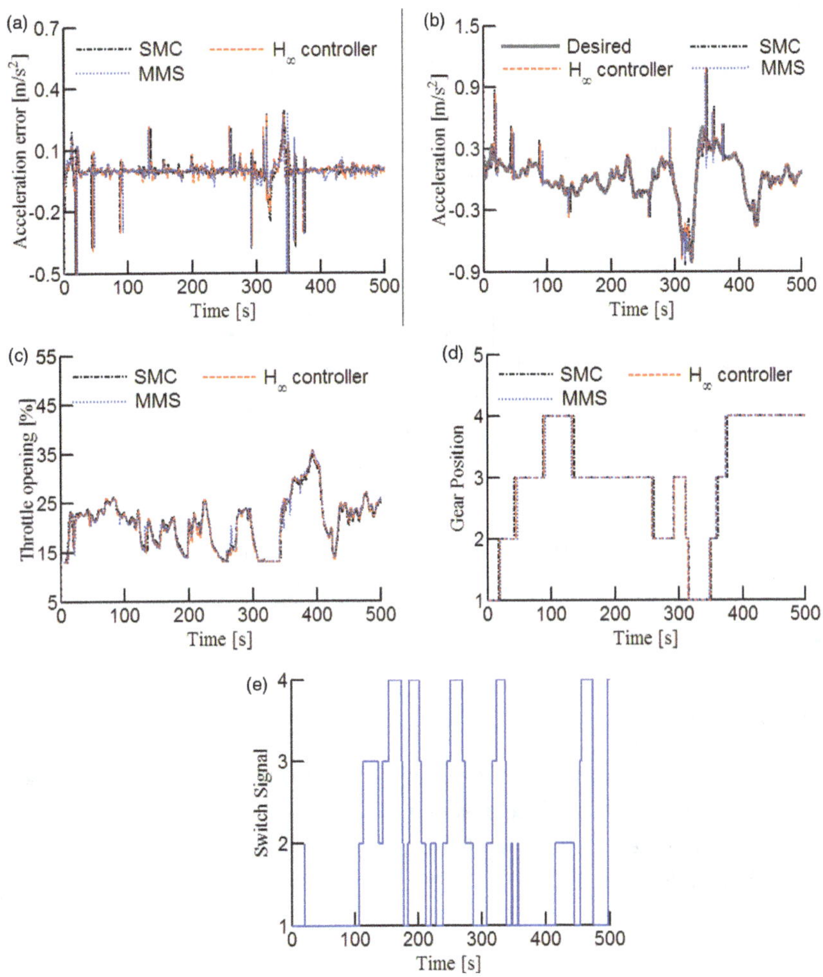

Figure 6. Results under normal condition. (a) Acceleration tracking error, (b) acceleration, (c) throttle control, (d) gear position and (e) switching signal.

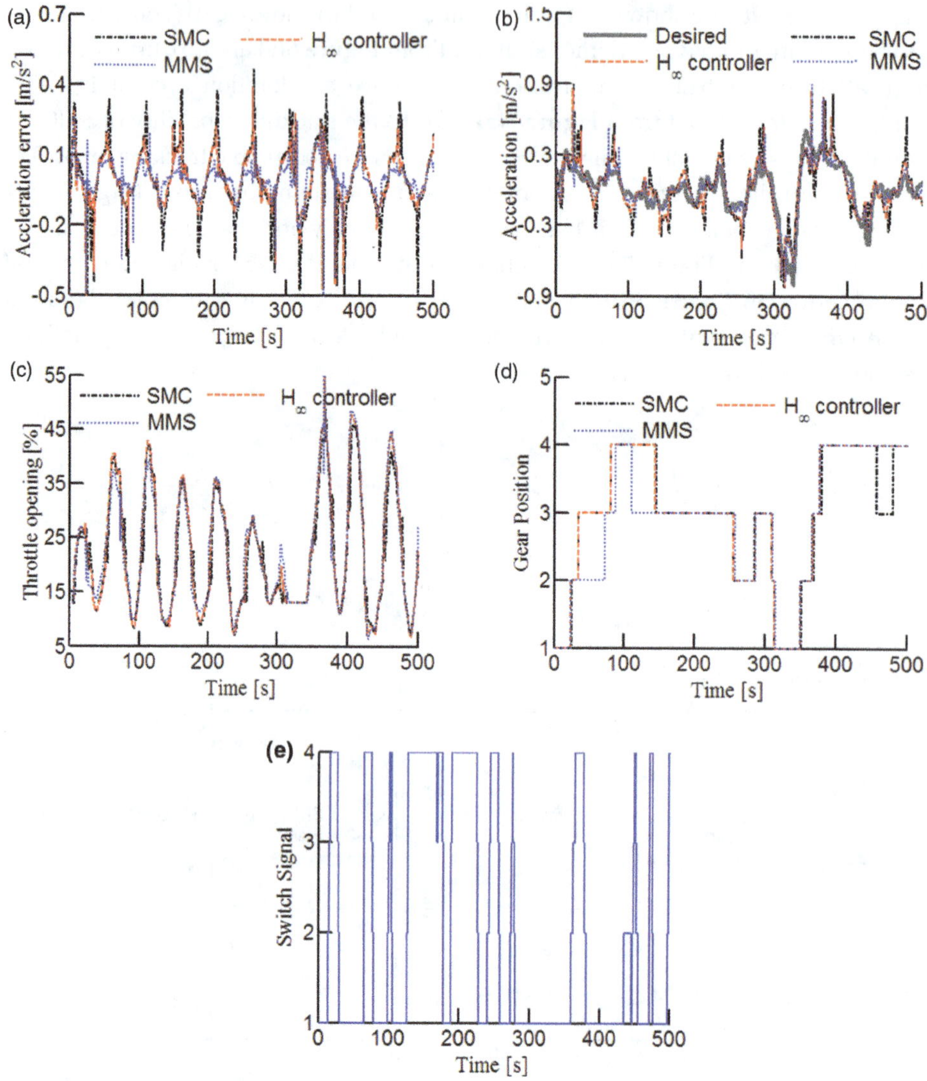

Figure 7. Results under disturbed condition. (a) Acceleration tracking error, (b) acceleration, (c) throttle control, (d) gear position and (e) switching signal.

A more deep simulation is conducted to analyse the influences of uncertain level on the tracking ability of acceleration and gear shifting behaviours. At this condition, the level of uncertainty is increased step by step and the relationship between the uncertain level and model uncertainties is described by:

$$
\begin{aligned}
M &= 1200\ \text{kg} + 40\text{kg}\cdot\varepsilon,\ \varphi_{\max} \\
&= 1\deg\cdot\varepsilon, v_{\text{wmax}} = 2\frac{\text{m}}{\text{s}}\cdot\varepsilon,
\end{aligned}
\tag{25}
$$

where ε represents the level of uncertainty, with maximum mass 1600 kg, maximum road slope 10 deg, and maximum wind speed 20 m/s (when $\varepsilon = 10$). $\varepsilon = 0$ implies that there is no model uncertainty. The root mean square error (RMSE) of acceleration is used to measure the capability of tracking. **Figure 8** presents the RMSE of acceleration error and the number of gear shifting per minute (denoted as N_{gear}/min). In nominal condition, the RMSE of acceleration error of the three robust controllers is almost the same. As the uncertainty level increases, the tracking capability of the SMC and H_∞ quickly drops, whereas the MMS still holds acceptable accuracy. **Figure 8(b)** is used to release the concern that the MMS might largely increase the number of gear shifting because of its switching structure.

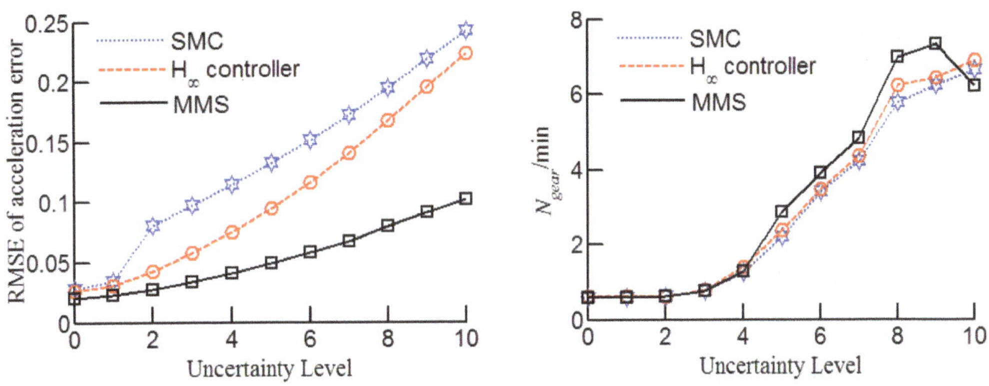

Figure 8. Performances under different uncertain levels.

5. Conclusions

This chapter proposes a robust acceleration control method for consistent node dynamics in a platoon of automated connected vehicles (CAVs). The design, which is based on multiple model switching (MMS) control structure, is able to offer more consistent and approximately linear node dynamics for upper level control design even under large uncertainties, including vehicle parametric variation, varying road slop and strong environmental wind. The following remarks are concluded:

(1) Homogeneous and linear node dynamics is important for platoon control. This requires the acceleration tracking performance to be accurate and consistent, and accordingly results in critical challenges because of the linearization error of powertrain dynamics and large model uncertainties in and around vehicles. The proposed MMS control structure can divide the large uncertainties of vehicle longitudinal dynamics into small ones. Accordingly, multiple robust controllers are designed from the multiple model set, and a scheduling logic is also presented to automatically select the most appropriate candidate controller into loop according to the errors between current vehicle dynamics and models.

(2) The designed switching index can measure the model error of vehicle longitudinal dynamics properly and the right acceleration controller is selected into the closed loop. The robust stability and performance of this acceleration tracking control system can be ensured. Both the simulation and experiment results demonstrate that this switching control system has better performances than that designed by either H_∞ control or sliding mode control approach in large uncertain conditions.

Acknowledgements

This was supported by the State Key Laboratory of Automotive Safety and Energy under Project No. KF16192 and NSF China with grant 51575293.

Author details

Feng Gao[1*], Shengbo Eben Li[2] and Keqiang Li[2]

*Address all correspondence to: gaofeng1@cqu.edu.cn

1 School of Electrical Engineering, Chongqing University, Chongqing, China

2 Department of Automotive Engineering, Tsinghua University, Beijing, China

References

[1] Luettel T, Himmelsbach M, Wuensche J. Autonomous ground vehicles—concepts and a path to the future. Proceedings of the IEEE 100. 2012; (Special Centennial Issue): 1831–1839. DOI: 10.1109/JPROC.2012.2189803.

[2] Shladover S. Longitudinal control of automotive vehicles in close-formation platoons. Journal of Dynamic Systems Measurement and Control. 1991; 113(2): 231–241. DOI: 10.1115/1.2896370.

[3] Sheikholeslam S, Desoer C. Longitudinal control of a platoon of vehicles with no communication of lead vehicle information: a system level study. IEEE Transactions on Vehicular Technology. 1993; 42(4): 546–554. DOI: 10.1109/25.260756.

[4] Hedrick J, Tomizuka M, Varaiya P. Control issues in automated highway systems. IEEE Transactions on Control System Technology. 1994; 14(6): 21–32. DOI: 10.1109/37.334412.

[5] Wang J, Rajamani R. Should adaptive cruise control systems be designed to maintain a constant time-gap between vehicles. IEEE Transactions on Vehicular Technology. 2004; 53(5): 1480–1490. DOI: 10.1109/TVT.2004.832386.

[6] Wang J, Longoria R. Coordinated and reconfigurable vehicle dynamics control. IEEE Transactions on Control System Technology. 2009; 17(3): 723–732. DOI: 10.1109/TCST. 2008.2002264.

[7] Gao F, Dang DF, Li SB. Control of a heterogeneous vehicular platoon with uniform communication delay. In: IEEE International Conference on Information and Automation. 8–10 Aug: Lijiang. 8–10 Aug: IEEE; 2015. p. 2419–2424. DOI: 10.1109/ICInfA. 2015.7279692.

[8] Stankovic S, Stanojevic M, Siljak D. Decentralized overlapping control of a platoon of vehicles. IEEE Transactions on Control System Technology. 2000; 8(5): 816–832. DOI: 10.1109/87.865854.

[9] Barooah P, Mehta P, Hespanha J. Mistuning-based control design to improve closed-loop stability margin of vehicular platoons. IEEE Transactions on Automatic Control. 2009; 54(9): 2100–2113. DOI: 10.1109/TAC.2009.2026934.

[10] Hao H, Barooah P. Stability and robustness of large platoons of vehicles with double-integrator models and nearest neighbor interaction. International Journal of Robust and Nonlinear Control. 2013; 23(18): 2097–2122. DOI: 1002/rnc.2872.

[11] Rajamani R, Shladover S. An experimental comparative study of autonomous and cooperative vehicle following control systems. Transportation Research Part C. 2001; 9(1): 15–31. DOI: 10.1016/S0968-090X(00)00021-8.

[12] Li ES, Li KJ, Wang JQ. Economy oriented vehicle adaptive cruise control with coordinating multiple objectives function. Vehicle System Dynamics. 2013; 51(1): 1–17. DOI: 10.1080/00423114.2012.708421.

[13] Gao F, Li XP, Ming GQ. Adaptive speed control under vehicle and road uncertainties using multiple model approach. In: American Control Conference; 4–6 June; Portland, OR. 4–6 June 2014: IEEE; 2014. p. 897–902. DOI: 10.1109/ACC.2014.6858598.

[14] Chen Z, Mi C, Xu J, Gong X, You C. Online energy management for a power-split plug-in hybrid electric vehicle based on dynamic programming and neural network. IEEE Transactions on Vehicular Technology. 2014; 63(4): 1567–1580. DOI: 10.1109/TVT. 2013.2287102.

[15] Fritz A, Schiehlen W. Automatic cruise control of a mechatronically steered vehicle convoy. Vehicle System Dynamics. 1999; 32: 331–344. DOI: 10.1076/vesd.32.4.331.2077.

[16] Fritz A, Schiehlen W. Nonlinear ACC in simulation and measurement. Vehicle System Dynamics. 2001; 36(2): 159–177. DOI: 10.1076/vesd.36.2.159.3556.

[17] Hunt K, Johansen T, Kalkkuhl J. Speed control design for an experimental vehicle using a generalized gain scheduling approach. IEEE Transactions on Control System Technology. 2000; 8(3): 381–295. DOI: 10.1109/87.845870.

[18] Li SB, Gao F, Cao DP, Li KQ. Multiple model switching control of vehicle longitudinal dynamics for platoon level automation. IEEE Transactions on Vehicular Technology. Forthcoming. DOI: 10.1109/TVT.2016.2541219.

[19] Swaroop D, Hedrick J, Choi S. Direct adaptive longitudinal control of vehicle platoons. IEEE Transactions on Vehicular Technology. 2001; 17(1): 150–161. DOI: 10.1109/CDC.1994.410877.

[20] Higashimata A, Adachi K. Design of a headway distance control system for ACC. JSAE Review. 2001; 22(1): 15–22. DOI: 10.1016/S0389-4304(00)00091-6.

[21] Yamamura Y, Seto Y. An ACC design method for achieving both string stability and ride comfort. Journal of System Design and Dynamics. 2008; 2(4): 979–990. DOI: 10.1299/jsdd.2.979.

[22] Gao F, Li KQ. Hierarchical switching control of longitudinal acceleration with large uncertainties. International Journal of Automotive Technology. 2007; 8(3): 351–359. DOI: 10.1109/ICVES.2006.371597.

[23] Xu Z, Ioannou P. Adaptive throttle control for speed tracking. Vehicle System Dynamics. 1994; 23: 293–306. DOI: 10.1080/00423119408969063.

[24] Keneth J, Tor A, Jens K. Speed control design for an experimental vehicle using a generalized scheduling approach. IEEE Transactions on Control System Technology. 2008; 8(3): 381–395. DOI: 10.1109/87.845870.

[25] Naranjo J, Gonzalez C, Carcia R. ACC+stop&go maneuvers with throttle and brake fuzzy control. IEEE Transactions on Intelligent Transportation System. 2006; 7(2): 213–225. DOI: 10.1109/TITS.2006.874723.

[26] Dai X, Li C, Rad A. An approach to tune fuzzy controllers based on reinforcement learning for autonomous vehicle control. IEEE Transactions on Intelligent Transportation System. 2005; 6(3): 285–293. DOI: 10.1109/TITS.2005.853698.

[27] Hespanha J, Liberzon D, Morse A. Hysteresis-based switching algorithms for supervisory control of uncertain systems. Automatica. 2003; 39: 263–272. DOI: 10.1016/S0005-1098(02)00241-8.

[28] Hespanha J, Liberzon D, Morse A. Multiple model adaptive control. Part 2: switching. International Journal of Robust and Nonlinear Control. 2001; 11(5): 479–496. DOI: 10.1002/rnc.594.

[29] Gao F, Li SB, Kum D, Zhang H. Synthesis of multiple model switching controllers using H∞ theory for systems with large uncertainties. Neurocomputing. 2015; 157: 118–124. DOI: 10.1016/j.neucom.2015.01.029.

[30] Bashivan P, Fatehi A. Improved switching for multiple model adaptive controller in noisy environment. Journal of Process Control. 2012; 22: 390–396. DOI: 10.1016/j.jprocont.2011.12.010.

[31] Kianfar R, Augusto B, Ebadighajari A. Design and experimental validation of a cooperative driving system in the grand cooperative driving challenge. IEEE Transactions on Intelligent Transportation System. 2012; 13(3): 994–1007. DOI: 10.1109/TITS.2012.2186513.

11

The Reliability of Autonomous Vehicles Under Collisions

Ibrahim Kutay Yilmazcoban

Abstract

Unmanned autonomous vehicles (UAV) subject is one of the most conspicuous topics of the last decade ensures a better handling ability for more precision way of control of the system or the drive compared to the human. Focusing this issue causes the more computerized vehicles the less unacceptable mistakes. Notwithstanding the efficient drive parameters are increased, the collisions are still the worst scenarios for the vehicles that should be taken care of. The most reliable way to define the collision response of the vehicles and the occupants is crash tests, although it is time consuming. Except crash test is the most convenient procedure to define the details of the impact process; it is also the most expensive way even for the major companies and research centers. Considering the compelling pricey amount of the first investment of the crashworthiness facilities and the crash test costs, recently, computer-aided simulations with the finite element method (FEM) using an explicit dynamics approach is very convenient, especially for the transient dynamic analysis. This chapter characterizes some perspectives of the autonomous vehicles in a short glance of an enormous science of the collisions with the help of experimental and numerical approaches.

Keywords: crash worthiness (impact), vehicle reliability, occupant safety, FEM, explicit dynamics, LSDYNA

1. Introduction

Over the decades, safety standards of the vehicles have been raised from the demands of the automobile users and organizations like Federal Motor Vehicles Safety Standards and Regulations (FMVSS) [1], National Highway Traffic Safety Administration (NHTSA) [2], European New Car Assessment Program (Euro-NCAP) [3], Society of Automotive Engineers (SAE) [4], etc. Superior technology evolved vehicles, which need more resources, offer diverse facilities for the occupants. Respecting the autonomous systems, more interventions of the drivers are starting to narrow. Although the yield of luxury and reliability, sometimes it operates against the occupants. During computer governing trips, individuals want to attain locations faster. In unexpected inaccuracy situations, automatic systems could cause fatal results.

However, the most common and important topic of vehicles is the crash-worthiness. Ensuring the reliability of the vehicles and safety of the occupants, the vehicle structure should be constructed in a way to absorb the impact effects and shock gradually, not to create a sudden deceleration which causes also injuries and fatalities. To decrease the failure risks, entire vehicle should be investigated, primarily starting from the basic structure of the vehicle under the perspectives of statics and strength of the materials. Then, the natural frequencies and the dynamic loading of the system should be incorporated into step by step. Generally for the dynamic loading, pre-captured dynamic road data also helps developing the vehicle structure. During the development period, beforehand defined systems like the concept design of the vehicle, and the shape should be protected, but the adaptation of the alterations should affect the design in minor.

Before determining the impact response and the behavior of the vehicle during a collision, the reasons of the accidents should be investigated. Serious amount of the accidents occur by the carelessness, negligence, and recklessness of the driver. The first one decreasing the risk of the collision is preventive driving techniques. This technique is a part of the education for the expert driving, although it refers time, patience, and talent. Yet, without any attention and enough reflexes, even this preventive method, results in frustration. In addition, if the all scenarios are considered the environmental and road conditions, and the faults of the other vehicles, including the drivers, effect the vehicles as other factors. Thence, the UAV technology becomes the redeemer of the system to solve the problems completely or decrease the errors to a sufficient level for a more reliable drive.

The most proper way to determine the causes is an experimental study, despite it is not a cost effective and time-consuming way. With the all other test methods such as positioning, loading, and calibrating, crash tests are the most significant ones. After the real-time collision tests, widely known experimental approach on the crash-worthiness researches is the sled test [2, 3, 5, 6]. The sled tests are more under-control when it is compared with the other methods. But again, the financial hassle is beyond the limits of the most companies and facilities.

Decreasing the expenses of the crash tests, many companies untangle their crash-worthiness difficulties after mandatory collision tests, with the computer-aided simulations. Consequently, computerized methods are immensely reliable, if almost, the last three decades are considered. To overcome the real-time simulations of the engineering problems, many developed computer-aided analysis softwares can be obtained from the market. Even though some of the softwares for an exact engineering solution such as structural analysis, impact analysis and thermal analysis remain part of the market is dealing with the multi discipline engineering solution packages serving for the finite element method (FEM). But for the general FEM and its solution methodology [7–9] is not focused on significant issues. Hence, the more applications in a software sometimes decreasing the expertise approach. According to our perspective, impact simulation codes indicate the prominent side of the vehicle collisions. Thereupon, for the impact procedures, explicit dynamics approach based on the FEM gets through the formidable problems, satisfactorily [10]. In **Figure 1**, FEM model (a) and a crash simulation (b) of a wheelchair and a seating system via sled scenario are given using the human mass adaptation to understand the crash worthiness [11].

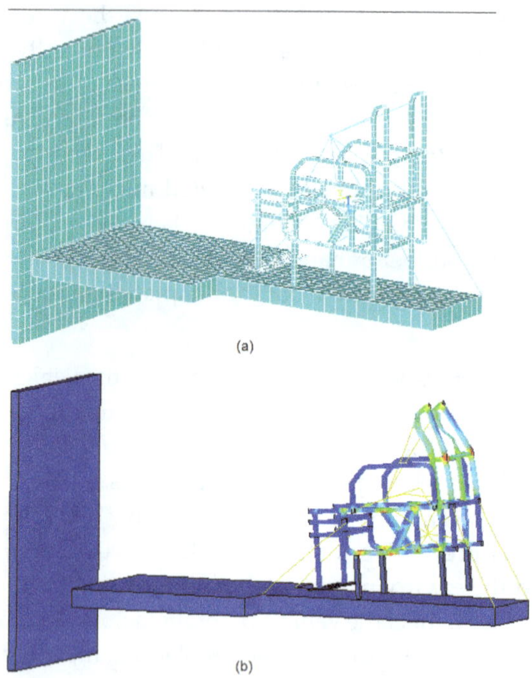

Figure 1. (a) Finite element model of a wheelchair crash simulation via sled test, (b) Finite element simulation result of the model for a defined specific time.

One of the most popular and extensively used software for the impacts is LSDYNA, the product of the Livermore Software Technology Corporation (LSTC) [12]. This code is used by some of the software companies as a fundamental explicit dynamics solver. With the help of the validated FEM results, reliable procedures will be determined to figure out the real vehicle crash scenarios in an inexpensive way, even for the UAV technology. During improvements invariably, reverse engineering applications are in the process from beginning till the end to mend the faults scores of times.

2. Autonomous vehicles: the perspective of collisions

Recently, the importance of the unmanned autonomous vehicles for every day, industrial and military applications are rapidly increasing, regarding the demand of the market. It is complex enough not to start the mass productions for today, especially manufacturing ability in excessive numbers of the automobile industry is considered. But, when the issue is saving a human being, we should reconsider everything referring the military developments against terrorism or wars. The defense industry focuses on the aerial vehicles to check the required zones and decrease the threats by the enemies or the terrorists. An unmanned aerial vehicle (UAV) shown in **Figure 2** in Ref. [13], widely used as a drone, or an unmanned aircraft system (UAS), and also referred by several other names, defines an aircraft without a human pilot. Therefore, an aircraft without a pilot means, alive human, but in the right hands yet, this sophisticated technology will be so perilous against innocents. And also, the more professionally UAS requires the more trained and talented personnel in particular formidable and arduous duties. This can possibly cause unexpected mistakes during the tasks that provoke impacts and collisions.

Figure 2. Medium Altitude Long Endurance (MALE) UAV of Turkish Aerospace Industries, Inc.

Figure 3. Google self-driving car.

The technological improvements in numerous dissimilar segments are effecting the UAV technology too, like media and entertainment. With the help of the lately used drones are easily operated, simple and cheap widely used by the professionals likewise amateurs in various applications. This option introduces some hazardous results with the ease of the system. Although the price of the simple applications like drones less when it is compared with the complex vehicles, consequences could be analogous.

If we focus on the automobile industry some leading companies like Google have been paying attention for autonomous cars for a while. See **Figure 3** [14]. After spending many years to research and development, the shown up products of UAV technology on the roads are a quite impressive story.

With regret, even for the finest tech. companies, some of the undesired incidents bring about the damaging results of the UAV. The self-driving car of the Google collided with a public transportation bus in California, USA (**Figure 4** in [15]) on February 14, 2016. In spite of the highly developed vehicle technology, it is indispensable to face with collisions for some cases.

Figure 4. Collided self-driving car of Google.

Whatever is done or planned for evolving the UAVs, sometimes these vehicles encounter with inevitable problems, and thus, it will be concluded with accidents as it was not planned. Under these circumstances, vehicle safety and crash worthiness should be taken care of in a comprehensive viewpoint.

3. Causes and prevention methods of collisions

Involuntary emerged incidents by the humans, vehicles, and environmental effects may cause accidents randomly. Sometimes, the reasons are known and the sufficient prevention methods can be created for the safety of the vehicles or the occupants or the both of the occupants and the vehicles. Although the vehicle is UAV, the opponents or the other vehicles could be driven by the humans. Thus, the human existence should be considered evermore, while the robustness of the vehicles is acceptable. With reference to known conversant reasons of the collisions, inaccurate parts of the mechanical, electronical, or software-based problems could be adjusted. However, the human originated mistakes which can differ from one being to another are much harder to fix with straightforward solution approaches. The psychological, physiological, behavioral potential, capability, and carefulness of the humans will alter the results of the happenings. Sometimes, the reasons are known, but it is not able be figured out by anything else the events like the several natural disasters. Otherwise, if the reasons are not specified or estimated for the collisions, the new methodologies must be evaluated to diminish the odds of failures for both humans and the vehicles.

Considering the vehicle, which is not a UAV, can be also investigated for the safety of the pedestrians, occupants, regular vehicles, and the autonomous vehicles. The idea of preventive driving technique or expert skills will come into the minds for the human driver. Initially, validity of this opinion could be reasonable with the existence of the natural aptitude alongside the education of the preventive driving. Although the careful driving is very important, the reflex of a human body is a prominent factor of the innate talent effecting throughout the trip. Sometimes, this gained time in milliseconds can save lives when the human body reacts against an unexpected incident. Reflexes can be improved with the help of the practice, while we can assume this, as a learned or educated reflex.

Otherwise, if the reflex response is not satisfying the sudden handling situations, preventive driving education avails to avert the collisions during or before the accidents. This education covers primarily the calm driving in contrast with the aggressive and fast driving, refraining from the instantaneous movements of the steering wheel conducts the vehicle, sudden acceleration and decelerations in particular in tough weather and environmental conditions. To ensure a safe trip, an admissible space between the vehicles is a necessary approach to acquiring an enough time along the duration of the collision. Subsequently, scanning all around the vehicle with eyes and computerized systems for every single threat is the key factor for a reliable trip, like a radar system to perceive the risky ventures which could be possibly induced by anything else.

Afterward, the driver will face with the most critical phase: *Decision*! And the worst scenario is *Uncertainty* as an opposite situation signifying that "the worst decision is better than a hesitant behaviour." If we consider the options for the decision:

- Deceleration,
- Pull over,
- Suddenly stop,
- Acceleration,
- Return,
- U turn,
- Slide,
- Skid,
- Drift,
- Changing lanes,
- Manage to do both of the selected ones or more, simultaneously or consecutively.

If there is no chance avoiding from the accident, try to be in the safe zone or select the least hazardous way via changing the speed and maneuvering the vehicle to save the lives and minimize the collision effects. Doing nothing and being petrified is not an option, it is just an ambivalent condition eventuating with an undesirable result.

Considering the hassle of the situation during collisions, these types of behaviors and applications are so difficult for a human to apply and need some expertise. It is not for everyone! Thus, if the computerized UAV system is capable to handle all of these situations at the same time, except some uncommon incidents, it will be much more robust and reliable for everyone, although we consider that sometimes human mind and body can create metaphysical and exceptional solutions.

4. Vehicle safety standards

When we talk about the Unmanned Autonomous Vehicles, it comprises the ground, aerospace, and even the naval vehicles. If we try to clarify the safety standards for all conditions, it will be in the detail redundantly. Ordinarily when the crash-worthiness and accidents are on the carpet, extensively known portrayal is covering the automobiles followed up by the aircrafts and the other vehicles.

For determining the vehicle safety standards of the automobiles, some major organizations take place in the decisive position in the industry and the market. However, the most important role in framing the ideology of the crash testing is in the hands of the Federal Motor Vehicle Safety Standards and Regulations (FMVSS) and the National Highway Traffic Safety Administration (NHTSA) in the USA as one of the leading countries for the vehicle safety. In the NHTSA internet site [16] of the US Department of Transportation for the Federal Motor Vehicle Safety Standards and Regulations, safety assurance defined in four major topics given below:

1. Crash avoidance: Standards between 101 and 135 (Part 571),
2. Crashworthiness: Standards between 201 and 224 (Part 571),
3. Post-crash standards: Standards as 301, 302, 303, 304 and 500 (Part 571),
4. Other regulations: Standards between 531 and 595.

It is very important to point out, especially the contents of the crashworthiness for our point of view:

- Occupant Protection in Interior Impact,
- Head Restraints,
- Impact Protection for the Driver from the Steering Control System,
- Steering Control Rearward Displacement,
- Glazing Materials,

- Door Locks and Door Retention Components,
- Seating Systems,
- Occupant Crash Protection,
- Seat Belt Assemblies,
- Seat Belt Assembly Anchorages,
- Windshield Mounting,
- Child Restraint Systems,
- Side Impact Protection,
- Roof Crush Resistance,
- Bus Emergency Exits and Window Retention and Release,
- Motorcycle Helmets,
- Windshield Zone Intrusion,
- School Bus Rollover Protection,
- School Bus Body Joint Strength,
- School Bus Passenger Seating and Crash Protection,
- Rear Impact Guards,
- Rear Impact Protection.

In another perspective for the crash testing, in the internet site of the crash network [17] expresses the headings in different groups: interior testing, dash board, frontal impact, side impact, steering wheel, seats, seat belts, rear impact, rollover, head rests, and roof crush. Considering the developments in the vehicle industry including automobiles and aircrafts, etc., especially the UAVs, many countries trying to improve new methodologies against the collisions comprising hundreds of standards. The major idea for the crash worthiness is protect the occupants or all beings, and eliminate or diminish the damages caused by the collisions.

5. Crash worthiness and crash tests

The occupant safety and vehicle reliability are the fundamental topics that should be developed against collisions for the crash worthiness. Although the technology is evolved in many fields especially for the materials science comprehending composite materials and withal super collapsible beams to absorb impact energy, again if we are not able to provide a safe drive for the occupant or the vehicle it is going to be a fact that can be concluded with the injuries or fatality for the driver, passengers or the people outside while the vehicle face with the physical damage like self-destruction as a result of collision to the wall, barrier or a structure; accident with another vehicle via front, behind or side impacts; and maybe a rollover incident by some other effects. These are by definition may differ from one incident to another defining the accidents of vehicles even for the autonomous systems. Whatever happens, the engineers should design a structure that can able to absorb some of the energy by collapsing itself or deforming to an adequate level to protect the human beings. If the structure is too rigid, there is an amplification occurs and affect everything in a poor way that is transferring the accelerations to the occupants causing worse situation. Therefore, even for the safety and reliability, optimum way is better to protect lives against accidents.

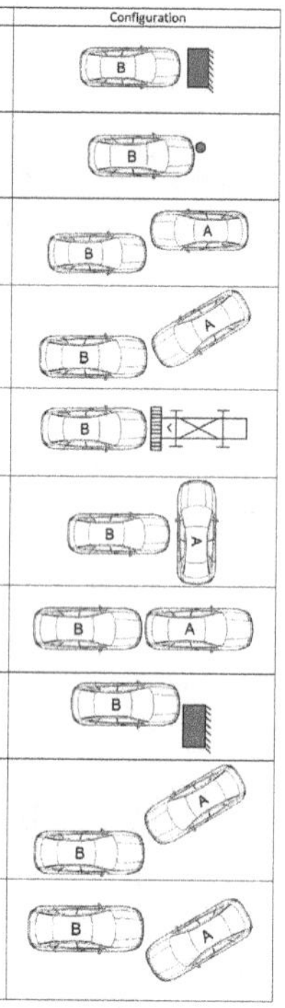

Actual Speed (Mph)	Type	Configuration
14.0 5.8 35.3 10.5 7.4	90 Barrier	
13.8 34.9	8" Pole Offset Left	
26.5	F-F 25% Overlap A ⟶ B	
25.2	30 L.F. – R.F. B ⟶ A	
35.3 35.4	90 Front ⟷ NHTSA MDB	
21.8 21.6	90 Front-Side (Both Moving)	
25.3 24.1	Front-Rear B ⟶ A	
17.7	Bumper Over-Ride 50% Offset Right	
27.7	30 L.F. Corner F.Center B ⟶ A	
37.5	30 R.F. Corner Side Wheel B ⟶ A	

Figure 5. Crash test modes.

To determine the crash response, broad amount of tests have been carried out for many years by the validated official agencies for crash worthiness. During the collisions under the combined loading conditions, the application of the crash energy management is dependent to plenty of situations, starting with contact mechanics and collapse modes that could be in the axial and/or in bending way causing many other reactions. According to Huang [5], the crash tests were chosen after reviewing some previous works around the world, accident statistics, and experience with the Ford Tempo air bag fleet based on the real-world crash investigations. The tests consisted of car-to-car tests versus barrier and fixed pole tests. Crash data were accumulated from the 22 vehicles; the chosen tests representing a vast range of accidents at or near the expected threshold of air bag deployment. There are 22 vehicles in 16 tests, as shown in (**Figure 5**), were used to gather the crash pulse data. For every test, the selected impact speed is a key factor to check the system performance of the air bag sensor will be activated or not during the incidents. Due to mentioned circumstances, we can define the types of the impacts [5]:

1. A perpendicular (90°) barrier,
2. A low- and high-speed rigid pole,
3. A front-to-front overlap,
4. An oblique front to front (three types),
5. A vehicle front to an moving deformable barrier (MDB) perpendicular,
6. A vehicle front-to-side,
7. A front-to-rear,
8. A bumper over-ride (or truck under-ride).

The tests can be classified in three categories: component tests, sled tests, and full-scale barrier impacts. Because of the complexity and many distinct test variables in full-scale tests, there is an acceptable reason in the reduction of the test's repeatability. The impact test depicted in (**Figure 6**) is representing a full-scale vehicle-to-vehicle impact test of Toyota Motor Company while they are developing a new safety system for their automobiles [18]. The dynamic and/or quasi-static response of an isolated component is defined by the component tests are exalted to detect the crush mode and energy absorbing ability. With the help of the component tests, it also gives the capability to the designers to develop prototype substructures and mathematical models. Moreover, the vehicle rear structure full-scale tests are governed either via a deformable barrier or a bullet car to state the integrity of the fuel tank. To evaluate roof strength according to FMVSS 216, a quasi-static load is applied on the "greenhouse," and providing that the roof deformation falls below a particular level for the applied load [6].

Figure 6. Vehicle-to-vehicle impact test of Totoya Motor Company.

For the sled test, engineers use a representation of a passenger compartment (buck) with all or some of the interior parts like seat, front panel, steering system, seat belts, and air bags. Anthropomorphic Test Devices or "Dummies" (ATD) or cadaver subjects are seated in the buck, utilized to simulate an occupant subjected to dynamic loads which is similar to a vehicle deceleration-time pulse, to determine the response in a frontal impact or side impact. To appraise the restraints of the system the sled test is a significant topic recorded by the high-speed cameras and photography applications for the dummy kinematics. It should also be stated that the performance of the ATD is determined at a higher impact speed of 35 mph in the NCAP test. Throughout the NCAP test, the restrain system is composed of three-point lap/shoulder belt system, and also additional restraint air bag. When we pay attention to the detail of the impact tests, numerous numbers of sensors as accelerometers, load cells, etc., located in the dummy and on the restraints, screen the forces and moments assisting to specify the impact reactions. Event data recorders have a very important role to understand the behavior of the system during accidents. However, when the recorded data will be analyzed digital filtering and reducing the noise (unwanted data) effect the results. Thus, it should be figured out cautiously.

Driving through the highways or freeways, the barriers are limiting the vehicles to get out of the road and protecting it, from crossings. During the barrier impact test there is a guided vehicle, canalized into a barrier with a predetermined initial velocity and angle. The purpose of the test is to maintain a structural integrity of the vehicle and adherence to the mandatory government regulations. The fully equipped vehicle collides a barrier at a 0°, +30°, and −30°, respectively, from an initial velocity of 30 mph.

The frontal offset impact test is another method acting with 40–50% overlap stating that the impact target could be rigid or deformable.

For the full-scale side impact test, a deformable barrier moving towards the left or right side of the vehicle with an initial speed at a certain angle (FMVSS 214). Especially for this test, side impact dummies ("SID" For the US and "EURO SID1" for Europe) which have different shape and properties, are used in the driver and outboard rear seat positions.

When the human parts are considered, the ATDs are widely used and capable enough to demonstrate the behaviors of the human nature in a fundamental way, even when it is indicating the spine, head, brain, organs, knee injuries, and whiplash incident in particular.

6. Explicit dynamics approach in FEM for the impact scenarios

The finite element method has been used by many scientists and engineers for decades to advance systems in different engineering applications for the validations, while reducing the experimental investments and the expenditure of the tests. This widely known method developed via the contributions of the researchers all around the world. While FEM problem is thriving, we will break the problem domain into a collection of 1D (dimensional), 2D and 3D shapes to form the discretization. Starting with the simplest way, $\{F\} = [K]\{D\}$ where F, K, and *D*, are the nodal force vector, global stiffness matrix and nodal displacement vector, respectively, as usually known. The displacements in this equation are the unknowns which should be found and lead us to figure out the strains and stresses, etc. To solve the equations, generally, the direct stiffness method, variational technique, or weighted residual approaches are used. However, this fundamental technique is now capable enough to handle the issues lately popular like the fracture mechanics with an extended finite element method, cohesive modeling, contact algorithms, smoothed-particle hydrodynamics (SPH) and Arbitrary Lagrangian–Eulerian Method (ALE), etc. Regarding recent a few decades, requisition of a nonlinear mechanics or engineering plasticity, for the delicate and slightly designs which should be robust in tough conditions, induced the durability tests including impacts and crashes. Explicit dynamics approach in FEM emerges with this idea to untangle the difficulties of the impacts, drop tests, explosions, and collision scenarios.

Vehicle technology under the perspective of the crash worthiness and safety during the collisions via crash tests to explicit dynamics simulations of the FEM is a trustworthy way to follow up. If we have a dekko to the literature some valuable studies can be determined [5, 6, 19] about the vehicle crash worthiness and the safety, while some books [6, 10, 19] explain the details of explicit dynamics one of the most challenging nonlinear problems in structural mechanics.

The implicit FE solution technique delivers accurate results as long at the stress levels in the beam do not exceed the yield-stress value, thus remaining in the linear (elastic) region of the material curve. Because the implicit solver usually obtains a result with a single step, according to the slope of the material curve, which is represented by Young's modulus, for the plastic region of the material there will be an inaccuracy whereat the slope is changed up the plastic characteristics. It means that the implicit solver is very powerful for linear (static) finite element analysis (FEA); thence, it provides an unquestioningly stable solution methodology.

Unfortunately, the implicit solution approximation may have difficulties with the immensely nonlinear finite element models. Additionally, inverting extremely large stiffness matrices with the implicit solver might consume too much time and memory. Although this method is not logical to use in highly nonlinear problems, it can able to calculate the equations with the help of the iteration cycle, which is applying the load in minor increments called as "timesteps" [19].

During the accident, the vehicle body faces with the impact loads propagate localized plastic hinges and buckling. So the large deformations and rotations with contact and heaping between the many parts can be determined. One of the most important ideas for the explicit dynamics is the wave effects, primarily involved in the deformation stage, associated with the higher stress levels. When the stress level exceeds the yield strength and/or the critical buckling load and the localized structural deformations comprise throughout a few wave transits in the frame of the vehicle. The effects of the inertia will chase this step and also dominate the ensuing transient reaction. The collision could be considered as a less dynamic incident if there is a comparison between the ballistic impact and the vehicle crash. Closed-form analytical problems for the impact scenarios in structural mechanics offer tough times for everyone. Therefore, the numerical methods arise with the practical options to decipher the unknowns.

A set of nonlinear partial differential equations of motion in the space-time domain is solved numerically by the explicit dynamics approximation, related to the stress–strain properties of the material about the initial and boundary conditions. To obtain a set of second order differential equations in time, the equations in space are discretized by the solution in the course of formulating the problem in a weak variational form and supposing an acceptable displacement field. Subsequently, in the time domain, discretization calculates the system of equations. If the selected integration parameters deliver the equations coupled, the technique is defined as implicit, while the solution is unconditionally stable. Furthermore, if the solution will be defined as explicit, the equations should be decoupled by the selection of the integration parameters, and it is conditionally stable [6].

As Du Bois et al. [6] described, a set of hyperbolic wave equations in the region of effectiveness of the wave front is figured out via the explicit dynamics, even then it does not require coupling of large numbers of equations. Yet, the unconditionally stable implicit solvers requiring assembly of a global stiffness matrix provide a solution for all coupled equations of motion. Especially considering the impact and vehicle crash simulations, including the utilization of contact, several material models and withal a combination of non-conventional elements, the explicit solvers show up more potent and computationally more productive than the implicit solvers, while the timestep is about two to three orders of magnitude of the explicit timestep. "Explicit" defines an equilibrium at a moment in time that the displacements of the whole spatial points are known in advance. With the help of the equilibrium, we can determine the accelerations; moreover, the use of central differencing technique assures to establish the displacements at the next timestep and reiterate the process. And also, the only inversion of the mass matrix, M, is enough for this procedure.

There should be stated that if the lumped-mass approximation is utilized the mass matrix will be diagonal, so there is no obligation of the matrix inversion. When the uncoupled equation system is figured out, the results will be in a very fast algorithm. For the minimum memory exigency, the assembly of a global stiffness matrix should be avoided by using an element-by-element approach for the calculation of the internal forces. The explicit integration will be second order accurate when it is applied attentively. Utilization of the related nodal displacements and/or velocities, the stresses are calculated discretely for each element, causing the effect of the loads on one part of the element onto the opposite part which is simulated by each timestep. Accordingly, the stress wave propagation is demonstrated through the element clearly. The deficiencies of the explicit solution technique are the conditional stability and the weak approximation of treating the static problems. The meaning of the conditional stability of the explicit integration algorithm is the integration timestep must be smaller than or equal to an upper bound value known as the Courant condition,: $\Delta t \le \frac{l_c}{c}$ explaining that the timestep must not pass over the smallest of all element timesteps identified by dividing the element characteristic length through the acoustic wave speed through the material from which origin the element is made like Du Bois et al. [6] mentioned. And they also said that the numerical timestep of the analysis must be smaller than, or equal to, the time required for the physical stress wave to pass by the element. If we try to explain it in a straightforward way, c of the mild steel elements is about 5000 m/s with a characteristic length of 5 mm for the automotive field practices concluding with 1 μs for the analysis timestep. Consequently, incidents dealing with the short duration of time are appropriate for the explicit dynamics solution technique enclosed with superior nonlinear behaviors and loading velocities demanding small timesteps for more precision sequels.

The integration of the equations of motion defines the time integration:

$$[M]\{\ddot{x}\}+[C]\{\dot{x}\}+\{F_{\text{int}}\}=\{F_{ext}\} \tag{1}$$

or in a different representation,

$$M\ddot{x}_n + C\dot{x}_n + f_{n_{\text{int}}} = f_{n_{ext}} \tag{2}$$

The basic problem is to determine the displacement, x_{n+1}, at time t_{n+1}. We can also say that $[K]\{x\} = \{F_{int}\}$. By the way, the explicit direct time integration can in a conceptual form be written as:

$$x_{n+1} = f(x_n, \dot{x}_n, \ddot{x}_n, x_{n-1}, \ldots) \tag{3}$$

The equation shows that the solution depends on nodal displacements, velocities, and accelerations at state n, quantities which are known. Therefore, the equation can be solved directly. And for the implicit:

$$x_{n+1} = f(\dot{x}_{n+1}, \ddot{x}_{n+1}, x_n, \dot{x}_n, \ldots) \tag{4}$$

This equation represents the solution that depends on nodal velocities and accelerations at state $n + 1$, quantities which are unknown. This implies that an iterative procedure is needed to compute the equilibrium at time t_{n+1}.

If we try to define the time increments:

$$\Delta t_{n+1/2} = t_{n+1} - t_n \tag{5}$$

$$t_{n+\frac{1}{2}} = \tfrac{1}{2}\left(t_{n+1} + t_n\right) \tag{6}$$

$$\Delta t_n = t_{n+\frac{1}{2}} - t_{n-\frac{1}{2}} \tag{7}$$

$$t_{n+1} = t_n - \Delta t_{n+\frac{1}{2}} \tag{8}$$

Then, the central difference method is used for the velocity:

$$\dot{x}_{n+\frac{1}{2}} = \frac{x_{n+1} - x_n}{\Delta t_{n+\frac{1}{2}}} \tag{9}$$

and

$$x_{n+1} = \dot{x}_{n+\frac{1}{2}} \Delta t_{n+\frac{1}{2}} + x_n \tag{10}$$

When $\Delta t_{n+\frac{1}{2}} = t_{n+1} - t_n$ is used, we get:

$$x_{n+1} = \dot{x}_{n+\frac{1}{2}} \left(t_{n+1} - t_n \right) + x_n \tag{11}$$

With the help of the acceleration, velocity can be stated as:

$$\ddot{x}_n = \frac{\dot{x}_{n+\frac{1}{2}} - \dot{x}_{n-\frac{1}{2}}}{t_{n+\frac{1}{2}} - t_{n-\frac{1}{2}}} \tag{12}$$

With $\Delta t_n = t_{n+\frac{1}{2}} - t_{n-\frac{1}{2}}$, the velocity is given below:

$$\dot{x}_{n+\frac{1}{2}} = \dot{x}_{n-\frac{1}{2}} + \Delta t_n a_n \tag{13}$$

From the Eq. (2), the acceleration at a_{n+1} can be found like:

$$\ddot{x}_{n+1} = m^{-1} \left(f_{n+1_{ext}} - f_{n+1_{\text{int}}} - C\dot{x}_{n+\frac{1}{2}} \right) \tag{14}$$

Remember, the damping force is taken at $\dot{x}_{n+\frac{1}{2}}$. The new velocity will be calculated with the old velocity and acceleration by Eq. (13). From this result, displacement will be updated at Eq. (10). Consequently, the new internal and external forces help for determination of the new acceleration [Eq. (14)].

Considering the LSDYNA, the widely known commercial explicit solver, the procedure for the solution steps is given below:

When $t = 0$, it will be the beginning of the solution loop or cycle. Whereupon, the displacement will be updated at Eq. (10), while the velocity is also updated at Eq. (13). Then, the internal forces (strain rate, strain, stress, and force magnitude) will be computed looping over the elements. Subsequently, external forces will be calculated to reach the computation of the accelerations at Eq. (14) which is finalizing the initial loop. Thereafter, following loops will go through the process again and again, until the convergence criteria will be satisfied.

For the robust design, stress wave propagation, timestep adjustment, critical element length and finally, the mesh size and quality are the factors, will be considered carefully to affect the simulation performance in a better way.

Due to the reliable side of the explicit dynamics solution methodology, we indicated the differences between the implicit and explicit solver for the approximation of the vehicle crash simulations and safety of the occupants and vehicles. We can possibly say that, for the collision of the regular or autonomous vehicles the accident results will be similar that should be taken care of precisely.

6.1. Crash simulations of the vehicles

The studies about the crash worthiness and impact scenarios as explained in the previous section are based on the explicit dynamics solution technique resulting with the satisfactory approximations. Some of the commercial codes for the explicit dynamics simulations are embedded inside the multidiscipline FEA programs such as ABAQUS, ANSYS, and MSC softwares, even though some of them working alone for just these types of applications as LSDYNA, PAMCRASH, etc. As it was mentioned, before these softwares are auxiliary engineering applications for solving the problems with the numerical methods. The most important task is the prediction of the failures of the system or structure of the vehicle and the problems that occupants can be faced. When the prediction of the simulations is validated, it can be said that FEM code has done its job properly.

To comprehend the validation of the simulations with the real crash tests, numerous topics should be investigated before, during or after the simulations. We can simply name these steps like: before the process is pre-processing, during is processing or simulation, and after the processing is post-processing.

Pre-process section is dealing with the creation of the finite element model, starting from the Computer-Aided Design (CAD) of the structure or the system, material descriptions or modelling, element selections, defining the boundary conditions (describing the loads, displacements and support properties, etc.) and initial values like velocities, adjusting the contact algorithms, total time and timestep size definition. For more elaboration, mass scaling factor and/or hourglass controls can be adjusted. Considering the information given, the geometry should be meshed as smooth shapes as possible. During the mesh generation element, sizes are so prominent, will affect the simulation time, according to the number of the elements. If the mesh is coarse, there will be a weak approximation and short period of processing time, notwithstanding if there is a finer meshed model exists, the convergence rate will be better, while the processing time is prolonged. Thus, there should be an optimum mesh size for the common structure; however, the significant zones are meshed with finer elements to increase the stability and for the better result.

When the simulation is processed, the finite element model is ready for the post-processing. The most considerable issue for the impact is absorb or get rid of the shock energy by different methods like geometrical change of the structure, material selection, and application. The second problem is the large deformation of the structural components. Additionally, changing the parameters of the simulation model will affect the results widely depending on the experience of the analyst. Because of dealing with the highly plastic problems and non-linearity during collisions, the analyst needs to possess deep knowledge about nonlinear finite element analysis for continuum mechanics [20–22]. It should be stated that the analyst can be faced with the geometrically nonlinear problems or material nonlinearities which is imported to be expressed.

Figure 7. The vehicle crash test and simulation.

Regarding the details of the post-processing, we will start to observe and evaluate the simulation outputs, step by step. While we are investigating the steps of the impact scenario, kinetic energy change, absorbed energy amount, deformations and displacements, strain and stress distribution, and some other major issues should be appraised attentively and the results should be validated with the real crash tests. Besides, the effects of the different components, deformation characteristics, and the changes of the parameters must be observed and the behavioral outcomes of the incident should be crosschecked with the real situation. The literature is replete with numerous of interesting researches about the crash-worthiness of the vehicles and the occupant safety including the Anthropomorphic Test Dummy (ATD) test and simulations [23–27]. Especially LSDYNA conferences are popular with the presentations of the explicit dynamics simulations and also most of the papers can be reachable from the internet site of "DynaLook" [28]. For a better representation of an explicit dynamics simulation, **Figure 7** demonstrates the full vehicle crash scenario [29].

7. Conclusion

Whole crash-worthiness studies are remarkable starting with the elementary functions of the profound methods of NCAP tests considering despite the costs of the experiments are enormous, contribution of the test results is challenging.

For a robust vehicle, the engineers should design a structure that can able to absorb some of the energy by collapsing itself or deforming to an adequate level to protect the human beings. If the structure is too rigid, there is an amplification occurs and affect everything in a poor way that is transferring the accelerations to the occupants causing worse situation. Therefore, even for the safety and reliability, optimum way is better to protect lives against accidents.

Nevertheless, for diminishing the costs of the studies of FEM and explicit dynamics methods, which are proceeding under the supervision of the computers and codes verified by the long-term researches and applications, are competent to handle the complicated algorithms yielding with satisfactory results. The incidents dealing with the short duration of time are appropriate for the explicit dynamics solution technique enclosed with superior nonlinear behaviors and loading velocities demanding small timesteps for more precision sequels.

Take into account the results of the experiments, analytic solutions and computerized methods are mentioned as FEM codes are robust approximations to understand the behavior of the vehicle and occupants in the course of statics, dynamics, and collisions. Consequently, the finite element method and explicit dynamics solutions have been proven to validate the experiments and real life experiences in the range between 80 and 100%.

It should be clearly stated that the autonomous vehicles decreasing the risks of the collisions when it is compared to the human driver controlled vehicle.

8. Nomenclature

8.1. Symbols and descriptions

c Acoustic wave speed through the material of which the element is made

[C] Damping matrix

{D} Nodal displacement vector

$\{F\}$ Nodal force vector or External load vector

$\{F_{ext}\}$ External forces

$\{F_{int}\}$ Internal forces

$[K]$ Global stiffness matrix

l_c Element characteristic length

[M] Lumped-mass matrix

n Indicates step n

x Displacement

$\dot{x}$ Velocity

$\ddot{x}$ Acceleration

t Time

Δt Timestep size

Author details

Ibrahim Kutay Yilmazcoban*

Address all correspondence to: kyilmaz@sakarya.edu.tr and kyilmazcoban@berkeley.edu

Department of Mechanical Engineering, University of California, Berkeley, CA, USA

Department of Mechanical Engineering, Sakarya University, Serdivan, Turkey

References

[1] Federal Motor Vehicle Safety Standards (FMVSS). FMVSS.COM [Internet]. Available from: http://fmvss.com/. Accessed: 2.27.16

[2] National Highway Traffic Safety Administration (NHTSA). NHTSA [Internet]. Available from: http://www.nhtsa.gov/. Accessed: 2.27.16

[3] European New Car Assessment Program (Euro-NCAP). The Official Site of The European New Car Assessment Programme [Internet]. Available from: http://www.euroncap.com/en. Accessed: 2.27.16

[4] Society of Automotive Engineers (SAE). SAE International [Internet]. Available from: http://www.sae.org/. Accessed: 2.27.16

[5] Huang M. Vehicle Crash Mechanics. USA: CRC Press LLC; 2002. 504 p. doi: 10.1201/9781420041866.ch2

[6] Du Bois P, Chou C C, Fileta B B, Khalil T B, King A I, Mahmood H F, Mertz H J, Wismans J. Vehicle Crashworthiness and Occupant Protection. MIchigan, USA: American Iron and Steel Institute; 2004. 388 p.

[7] Logan D L. A First Course in the Finite Element Method. 5th ed. Stamford, CT, USA: Cengage Learning; 2012. 1000 p.

[8] Bathe K J. Finite Element Procedures in Engineering Analysis. New Jersey, USA: Prentice Hall; 1982. 735 p.

[9] Moaveni S. Finite Element Analysis Theory and Application with ANSYS. New Jersey, USA: Prentice Hall; 1999. 527 p.

[10] Wu S R, Gu L. Introduction to the Explicit Finite Element Method for Nonlinear Transient Dynamics. Hoboken, New Jersey, USA and Canada: WILEY, A John Wiley & Sons, Inc.; 2012. 352 p. doi:10.1002/9781118382011.ch1

[11] Yilmazcoban I K. Analysis of Wheelchairs' Frame Soundness under Environmental Conditions: Use of Finite Element Technique [thesis]. Sakarya, Turkey: Sakarya University; 2009. 162 p.

[12] Livermore Technology Software Corporation. Welcome to Livermore Software Technology Corp. [Internet]. 3.5.16 [Updated: 3.5.16]. Available from: http://www.lstc.com/. Accessed: 3.25.16

[13] TAI (Turkish Aerospace Industries, Inc.). Medium Altitude Long Endurance (MALE) UAV [Internet]. 3.31.2016 [Updated: 3.31.2016]. Available from: https://www.tai.com.tr/en. Accessed: 3.31.2016

[14] Google. Google Self-Driving Car Project [Internet]. 3.31.2016 [Updated: 3.31.2016]. Available from: https://www.google.com/selfdrivingcar/. Accessed: 3.31.2016

[15] HNGN (Headlines & Global News). Google Shares Blame For Autonomous Car Collision [VIDEO] [Internet]. 3.31.2016 [Updated: 3.31.2016]. Available from: http://www.hngn.com/articles/187103/20160310/google-shares-blame-v-day-autonomous-car-collision-video.htm. Accessed: 3.31.2016

[16] NHTSA. Federal Motor Vehicle Safety Standards and Regulations (Safety Assurance) [Internet]. 04.06.2016 [Updated: 04.06.2016]. Available from: http://www.nhtsa.gov/cars/rules/import/FMVSS/. Accessed: 04.06.2016

[17] Crash-Network (The web portal for the crash test engineer). FMVSS Standards related to Crash Testing [Internet]. 04.06.2016 [Updated: 04.06.2016]. Available from: http://www.crash-network.com/Regulations/FMVSS/fmvss.html. Accessed: 04.06.2016

[18] The Japan Times News. Imperial Family's car woes sparked Toyota whistleblower [Internet]. 04.17.2014 [Updated: 04.17.2014]. Available from: http://www.japantimes.co.jp/news/2013/06/09/business/corporate-business/imperial-familys-car-woes-sparked-toyota-whistleblower/#.VxPdUMJf2Ul. Accessed: 04.17.2014

[19] Christensen J, Bastien C. Nonlinear Optimization of Vehicle Safety Structures: Modeling of Structures Subjected to Large Deformations. 1st ed. US: Elsevier, Butterworth-Heinemann; 2016. 474 p.

[20] Belytschko T, Liu W K, Moran B, Elkhodary K I. Nonlinear Finite Elements for Continua and Structures. 2nd ed. UK: Wiley; 2014. 834 p.

[21] Reddy J N. An Introduction to Nonlinear Finite Element Analysis. 2nd ed. Great Britain: Oxford University Press; 2015. 687 p. doi:10.1093/acprof:oso/9780198525295.001.0001.

[22] Borst R, Crisfield M A, Remmers J J C, Verhoosel C V. Non-linear finite element analysis of solids and structures. 2nd ed. UK: Wiley; 2012. 542 p.

[23] Bambach M R. Fibre composite strengthening of thin-walled steel vehicle crush tubes for frontal collision energy absorption. Thin-Walled Structures. 2013;66:15–22. doi: 10.1016/j.tws.2013.02.006

[24] Golman A J, Danelson K A, Miller L E, Stitzel J D. Injury prediction in a side impact crash using human body model simulation. Accident Analysis and Prevention. 2014;64:1–8. doi:10.1016/j.aap.2013.10.026

[25] Al-Thairy H, Wang Y C. Simplified FE vehicle model for assessing the vulnerability of axially compressed steel columns against vehicle frontal impact. Journal of Constructional Steel Research. 2014;102:190–203. doi:10.1016/j.jcsr.2014.07.005

[26] Liu W, Zhao H, Li K, Su S, Fan X, Yin Z. Study on pedestrian thorax injury in vehicle-to-pedestrian collisions using finite element analysis. Chinese Journal of Traumatology. 2015;18:74–80. doi:10.1016/j.cjtee.2015.03.003

[27] Crocetta G, Piantini S, Pierini M, Simms C. The influence of vehicle front-end design on pedestrian ground impact. Accident Analysis and Prevention. 2015;79:56–69. doi: 10.1016/j.aap.2015.03.009

[28] DYNALook. Papers from European and International LS-DYNA User Conferences [Internet]. 04.14.2016 [Updated: 04.14.2016]. Available from: http://www.dynalook.com/. Accessed: 04.14.2016

[29] Structural crash. Esi-Get it right [Internet]. 04.17.2016 [Updated: 04.17.2016]. Available from: http://virtualperformance.esi-group.com/?q=applications-structural-crash. Accessed: 04.17.2016

Permissions

All chapters in this book were first published by InTech Open; hereby published with permission under the Creative Commons Attribution License or equivalent. Every chapter published in this book has been scrutinized by our experts. Their significance has been extensively debated. The topics covered herein carry significant findings which will fuel the growth of the discipline. They may even be implemented as practical applications or may be referred to as a beginning point for another development.

The contributors of this book come from diverse backgrounds, making this book a truly international effort. This book will bring forth new frontiers with its revolutionizing research information and detailed analysis of the nascent developments around the world.

We would like to thank all the contributing authors for lending their expertise to make the book truly unique. They have played a crucial role in the development of this book. Without their invaluable contributions this book wouldn't have been possible. They have made vital efforts to compile up to date information on the varied aspects of this subject to make this book a valuable addition to the collection of many professionals and students.

This book was conceptualized with the vision of imparting up-to-date information and advanced data in this field. To ensure the same, a matchless editorial board was set up. Every individual on the board went through rigorous rounds of assessment to prove their worth. After which they invested a large part of their time researching and compiling the most relevant data for our readers.

The editorial board has been involved in producing this book since its inception. They have spent rigorous hours researching and exploring the diverse topics which have resulted in the successful publishing of this book. They have passed on their knowledge of decades through this book. To expedite this challenging task, the publisher supported the team at every step. A small team of assistant editors was also appointed to further simplify the editing procedure and attain best results for the readers.

Apart from the editorial board, the designing team has also invested a significant amount of their time in understanding the subject and creating the most relevant covers. They scrutinized every image to scout for the most suitable representation of the subject and create an appropriate cover for the book.

The publishing team has been an ardent support to the editorial, designing and production team. Their endless efforts to recruit the best for this project, has resulted in the accomplishment of this book. They are a veteran in the field of academics and their pool of knowledge is as vast as their experience in printing. Their expertise and guidance has proved useful at every step. Their uncompromising quality standards have made this book an exceptional effort. Their encouragement from time to time has been an inspiration for everyone.

The publisher and the editorial board hope that this book will prove to be a valuable piece of knowledge for researchers, students, practitioners and scholars across the globe.

List of Contributors

Cheng Siong Chin
Newcastle University, Singapore, Singapore

Xionghu Zhong
Visenti Pte Ltd, Singapore, Singapore

Rongxin Cui
School of Electronic and Control Engineering, Northwestern Polytechnical University, Chang'an University, Xi'an, China

Chenguang Yang
Zienkiewicz Centre for Computational Engineering, Swansea University, Swansea, UK

Mohan Venkateshkumar
Department of EEE, Aarupadai Veedu Institute of Technology, Chennai, Tamil Nadu, India

Chika Yinka-Banjo and Olasupo Ajayi
University of Lagos, Lagos, Nigeria

Jing Xin, Xing Du, Yaqian Shi and Ding Liu
Shaanxi Key Laboratory of Complex System Control and Intelligent Information Processing, Xi'an University of Technology, Xi'an, P.R. China

Jian Zhang
Faculty of Engineering and Information Technology, University of Technology (UTS), Sydney, Australia

Mustafa Serdar Genç, Kemal Koca and Halil Hakan Açıkel
Wind Engineering and Aerodynamic Research Center, Department of Energy Systems Engineering, Erciyes University, Kayseri, Turkey

Sheng-Chih Shen and Yi-Ting Yang
Department of Systems and Naval Mechatronic Engineering, National Cheng Kung University, Tainan City, Taiwan ROC

Hacımurat Demir
Wind Engineering and Aerodynamic Research Center, Department of Energy Systems Engineering, Erciyes University, Kayseri, Turkey
Department of Mechanical Engineering, Aksaray University, Aksaray, Turkey

Mathias Fassini Mantelli
Federal University of Rio Grande do Sul, Porto Alegre, Brazil

Kondeti Lakshmi Vasudev
Department of Mechanical Engineering, Hindustan Institute of Technology and Science, Chennai, India

Ayodele Abiola Periola
Department of Electrical, Electronics and Computer Engineering, Bells University of Technology, Ogun State, Nigeria

Timothy Sands
Department of Mechanical Engineering, Stanford University, Stanford, USA

Kevin Bollino
Department of Electrical and Computer Engineering, George Mason University, Fairfax, USA

Feng Gao
School of Electrical Engineering, Chongqing University, Chongqing, China

Shengbo Eben Li and Keqiang Li
Department of Automotive Engineering, Tsinghua University, Beijing, China

Ibrahim Kutay Yilmazcoban
Department of Mechanical Engineering, University of California, Berkeley, CA, USA
Department of Mechanical Engineering, Sakarya University, Serdivan, Turkey

Index

www.ingramcontent.com/pod-product-compliance
Lightning Source LLC
LaVergne TN
LVHW080634210926
848485LV00003B/258
* 9 7 8 1 6 4 7 2 6 6 0 1 1 *